Arduino for Radio Amateur Applications

Program and build Arduino-based ham station utilities, tools, and instruments

Glen Popiel, KW5GP

elektor
design > share > earn

This is an Elektor Publication. Elektor is the media brand of
Elektor International Media B.V.
PO Box 11, NL-6114-ZG Susteren, The Netherlands
Phone: +31 46 4389444

Declaration

British Library Cataloguing in Publication Data
A catalogue record for this book is available from the British Library

ISBN 978-3-89576-601-5 Print
ISBN 978-3-89576-602-2 eBook

www.elektor.com
Editor: Jan Buiting, MA
Prepress Production: D-Vision, Julian van den Berg
Print: Ipskamp Printing, Enschede, The Netherlands

Contents

About The Author

Glen Popiel, KW5GP, is a retired network engineer and technology consultant and the author of the ARRL's Arduino for Ham Radio series of books, High Speed Multimedia for Amateur Radio, and numerous magazine articles and product reviews. He is also a regular on W5KUB.com's Amateur Radio Roundtable weekly webcast.

Always taking things apart (and sometimes even getting them to work afterwards), he discovered electronics in high school and has never looked back. As a teenager, he had one of the first "home computers", a Digital Equipment (DEC) PDP-8 minicomputer, complete with state-of-the-art Model 35 Teletype, in his bedroom that he and his friends salvaged from the scrap heap. Over his 50-year career, he has worked for various aerospace and computer manufacturers on radio and military turbojet research data acquisition and control systems, as well as hospital data systems, large-scale governmental agency networks, and utility company data networks.

First published in Kilobaud Microcomputing in 1979 for a project he designed using the RCA CDP1802 "Cosmac" microprocessor, he continues to work with microcontrollers and their applications in Amateur Radio. Since discovering the Arduino a number of years ago, he has developed a passion for this small, powerful, and inexpensive microcontroller and has given a number of seminars and hamfest forums on the subject of microcontrollers and Amateur Radio.

He is a member of the Olive Branch Amateur Radio Club (OBARC), Chickasaw Amateur Radio Association (CARA), and QRP Amateur Radio International (QRP-ARCI). Glen is also a former cat show judge and has exhibited his champion Maine Coon cats all over the United States, with the highlight of winning a Best in Show at Madison Square Garden in 1989.

He currently resides in Southaven, Mississippi, where he continues to create fun and exciting new Arduino projects for ham radio with his mischievous Maine Coon companions, Shadow and Angel.

Acknowledgements

This book is dedicated to my father, who has always encouraged and inspired me to learn and try new things. Without his guidance and support, I would never have discovered my love for electronics and ham radio.

My personal Arduino adventure would never have happened if my friend, Tim Billingsley (SK), KD5CKP, had not introduced me to the Arduino in 2011. Tim was my sounding board, spellchecker, and grammar coach for my first Arduino book, and somehow managed to keep me sane throughout the whole process. I also need to thank Craig Behrens (SK), NM4T, for his knowledge, guidance, and support in helping my first book become a reality. There are too many others who have helped me along the way to list here, but I want to take this opportunity to thank them all as well.

How This Book is Organized

This book contains a collection of amateur radio-related projects for the Arduino microcontroller. While the projects in this book were designed with the Arduino Nano in mind, you can also substitute an Arduino UNO without having to make any additional design changes.

While I understand your desire to skip ahead and dive right into the projects, I urge you to take the time to read the chapters leading up to the projects. These preliminary chapters will provide you with vital information and insights into the various modules, components, and technologies used in the projects that follow. They will also provide you with information and guidance on creating a good, functional work area to build and troubleshoot your Arduino projects. The more prepared and organized you are in these areas, the better things will often turn out. This doesn't necessarily mean your work area will stay clean and uncluttered for very long, but it helps to at least start out that way. After working for 20 minutes on a project, my work area tends to degenerate into a complete and total mess, but at least I had good intentions when I started.

Chapter 1, "Welcome to the Arduino", introduces you to the Arduino microcontroller and its history, and also provides a basic understanding of the concepts of Open Source as it pertains to the Arduino, along with the various types of Open Source licenses.

Chapter 2, "Arduino Boards and Variants", discusses the most common Arduino processor boards with an emphasis on the newer Arduino and Arduino-compatible boards, highlighting the boards of interest for use in ham radio-related projects.

Chapter 3, "Arduino Modules and Shields", discusses the various boards and modules that can be used to interface with the Arduino, allowing the Arduino to sense and communicate with the outside world, again, with an emphasis placed on those devices that lend themselves for use in ham radio-related projects.

Chapter 4, "Creating Sketches and Documenting Arduino Projects", introduces the Arduino Integrated Development Environment (IDE), creating sketches (programs), designing and flowcharting sketches, and documenting your Arduino projects.

Chapter 5, "Project Ideas, Tools, Construction, and Troubleshooting", discusses the project creation process, including how to decide what projects to build. It also covers tools, test equipment, and methods on how to troubleshoot your sketches, library problems, hardware, and other issues you may encounter while building Arduino projects.

Chapter 6, "Arduino I/O Methods", discusses in detail the I/O capabilities of the Arduino, including some of the newer methods now available, which method is best for communicating with the various shields and components, and how to best implement each I/O method.

Chapter 7, "Project 1 — CW Beacon and Foxhunt Keyer", shows how to build a CW transmitter beacon controller and foxhunt keyer.

Chapter 8, "Project 2 — Weather Station", covers how to build a mini-weather station that can sense and display temperature, humidity, barometric pressure, and wind speed without any moving parts.

Chapter 9, "Project 3 — RF Probe with LED Bar Graph", measures RF signal strength and displays it using an addressable RGB LED strip.

Chapter 10, "Project 4 — Touch-tone Encoder", shows how to build a 16-button touch-tone encoder to control devices remotely.

Chapter 11, "Project 5 — Touch-tone Decoder", shows how to build a 16-button touch-tone decoder to control devices remotely.

Chapter 12, "Project 6 — Waveform Generator", shows how to use a programmable Direct Digital Synthesis (DDS) module to generate sine, square, and triangle waves.

Chapter 13, "Project 7 — Auto Power On/Off", shows how to build an automatic power on/off project to automatically power on and turn off devices in your vehicle based on whether the engine is running or not, including programmable startup and shutdown delays.

Chapter 14, "Project 8 — Bluetooth CW Keyer", shows how to wirelessly connect your CW keyer paddle to your transmitter via Bluetooth.

Chapter 15, "Project 9 — Station Power Monitor", shows how to monitor your shack's DC power supply voltage and current, as well as ambient temperature.

Chapter 16, "Project 10 —AC Current Monitor", shows how to non-intrusively measure AC current, calculate the Kilowatt/hour usage, and the estimated cost of the measured power usage.

Chapter 17, "In Conclusion", discusses Arduino processors, devices, projects, ideas, and concepts not included in this book to provide inspiration and ideas for other projects to take you beyond the scope of this book.

Introduction

Welcome to Arduino for Amateur Radio. I have often said that working with the Arduino is more of an adventure than anything else. Based on my own personal experiences with the Arduino, nothing could be closer to the truth. When I started out on my own Arduino adventure a little over 12 years ago, my first thoughts were of all of the amateur radio projects I could build with the Arduino. Previously, such projects had seemed too complex or expensive to build. At the same time, I did not want to "reinvent the wheel" and revisit amateur radio projects that had already been created just for the sake of using an Arduino, but rather focus on new and unique projects ideas that would be ideally suited for the Arduino.

Additionally, I wanted each of my projects to highlight the unique features of the Arduino, the various hardware modules available, and go beyond the introductory "Hello World" and 'blinking lights" projects so often associated with basic Arduino projects. Emphasis was placed on creating a wide range of Arduino-based amateur radio projects that could easily be constructed in just a few short days. Along the way, I found that certain modules and components were simply not well documented or understood. I've done my best to "demystify" and simplify the usage of these modules and components so that you can easily incorporate them into your Arduino projects.

My goal was to provide ham-related Arduino projects that would also serve as a platform for you to not only learn about the Arduino, but also to allow you to expand upon the projects themselves, adding your own personal touches to your Arduino projects. To help you on this journey, this book starts by introducing some of the more common Arduino processor boards and add-on components that I have found to be most useful in creating my amateur radio Arduino projects. Each of these boards or components are described in detail, providing you with valuable information about the components that are available for you to use in your own Arduino amateur radio projects.

Above all, don't be afraid to experiment and try something new. The Arduino is inexpensive and easy to work with. The price of Arduino processor boards and modules tends to take away the fear of making mistakes and damaging the board. Without that fear, it's so much easier to learn how to solder and build projects, and not obsess over what happens if you make a mistake. Go for it and have fun along the way.

Since there are a large number of introductory Arduino books already available, this book assumes that you have a working knowledge of the Arduino and the Arduino Integrated Development Environment (IDE). If you are new to the Arduino, there are several excellent books and tutorials that you may find helpful. My personal favorites include *Beginning Arduino* by Michael McRoberts (ISBN 978-1430232407) and *Arduino Cookbook* by Michael Margolis (ISBN 978-1449313876). Another excellent book for learning the hardware interfacing aspect of the Arduino is *Exploring Arduino: Tools and Techniques for Engineering Wizardry* by Jeremy Blum (ISBN 978-1-118-54936-0). There are also some excellent Arduino tutorials available online at www.arduino.cc, www.learn.adafruit.com, and https://learn.sparkfun.com. Recently, I have discovered several other very good online course and tutorials available for the Arduino. The Programming Learning Academy (www.programmingelectronics.com) offers a free 12-lesson Arduino Video Course and an inexpensive e-book PDF download of the *Arduino Course for Absolute Beginners* by Michael James. Both are excellent Arduino learning tools. YouTube also has a tremendous amount of tutorial videos for the Arduino.

This book also assumes that you have a basic working knowledge of electronic components and construction techniques. Because the Arduino is so easy to work with, you don't have to be an expert, but you should be able to solder and build basic electronic projects. If you're new to electronics, check out *Electronics For Dummies* by Cathleen Shamieh (ISBN 978 1119675594), *Understanding Basic Electronics 2nd Edition* by Walter Banzhaf, WB1ANE (ISBN 978-087259-082-3), *Beginner's Guide to Reading Schematics* by Stan Gibilsco (ISBN 978-1260031102), and the *ARRL Handbook* (ISBN 978-1-62595-158-8), published annually by the American Radio Relay League (ARRL). Personally, I feel that every ham should have a copy of the *ARRL Handbook* in their library. I find myself constantly referring to mine for research or when I want to learn some new aspect of electronics or ham radio.

Since the projects in this book are ham radio-related, you may need an Amateur Radio Operator's license to use those projects on the air. While not all of the projects require a ham license, I strongly encourage you to become a ham if you are not already. As a ham, you'll find that there is strength in numbers, and more than likely, you can find other hams in your area that are interested in constructing Arduino and other projects. There is something for everyone in the ham radio community, and there is no reason for you to go it alone as you begin your Arduino adventure. I recommend that you find a local club at www.arrl.org/find-a-club and attend a meeting or two. You will not find a friendlier, more helpful group of people anywhere, and there's a good chance that you'll find other Arduino enthusiasts to work with. If you're interested in becoming a radio amateur, check out *Ham Radio for Dummies* (ISBN 978-1119695608) by H. Ward Silver, N0AX, and the *ARRL Ham Radio Licensing Manual.*

All of the sketches in this book were created using **version 1.8.19** of the Arduino IDE available from www.arduino.cc. The libraries and sketches for the projects in this book are available from www.kw5gp.com/Elektor. If you encounter any problems compiling the sketches, please verify that you are using the correct version of the IDE and that you have the libraries properly installed. And by all means, feel free to contact me, I'll be glad to help as much as I can.

Glen Popiel - KW5GP

kw5gp@arrl.net

Fall 2023

Chapter 1 • Welcome To The Arduino

Introduction

Since its introduction in 2005, the Arduino has become wildly popular among the hobbyist community. There are now an estimated ten million Arduino UNO boards in use, not counting the many "clone" and "variant" boards produced under the Arduino's unique Open Source licensing model. With its standalone single-board design, the Arduino can be interfaced to a wide variety of sensors and devices easily and inexpensively. Originally based on the Atmel (now Microchip Technology) series of microcontrollers, the Arduino, has onboard digital and analog I/O pins, along with support for two industry-standard bus protocols, Serial Peripheral Interface (SPI) and Inter-Integrated Circuit (I^2C). This built-in functionality is what helps make the Arduino such an easy and inexpensive way to build powerful and versatile electronic projects.

Released under the Open Source Creative Commons Attribution Share-Alike license, the Arduino is totally Open Source. From the board designs and schematic files to the program sketches (a sketch is the term for an Arduino program) and libraries, everything is Open Source, and you are free to do whatever you desire, as long as you properly attribute the authors in your work and share any changes you make to the existing code and libraries. For the most part, this means that everything about the Arduino is either free or very low cost. Because of this Open Source freedom to develop and market Arduino–compatible boards, the term "Arduino" is often used to describe both the official Arduino boards, as well as the many compatible boards. Over time, the term "Arduino" has also become the generic designation for the many other microcontrollers that can be used with the Arduino Integrated Development Environment (IDE).

One of the primary benefits of Open Source is that you have a whole community of hobbyists developing and freely sharing their projects. This can save you many hours of work if someone is working on a project similar to yours. You can integrate their libraries and code into your project, turning what could have been a months-long programming ordeal into a much shorter, more enjoyable path to a finished project.

Along with the Arduino board itself, there is a vast selection of components and modules designed to interface with the Arduino. These devices use the various device communication protocols such as SPI and I^2C that are already built into the Arduino, allowing simple connections to the Arduino using only a few wires. Now you can create complex projects without having to dig through piles of datasheets and solder for months like you had to in days gone by. Through the use of pre-existing "libraries" that are used to communicate with these external modules, often all that is needed is a little bit of additional coding to create a fully functional project. Libraries are pre-written blocks of code, similar to the device drivers you install on a PC, that you can use to interface external modules such as a TFT (Thin-Film Transistor) display without having to know the technical details of how to communicate with the device. Additional libraries are easily added to the IDE, allowing instant access to everything you need to integrate a new module or device into your Arduino project.

Figure 1.1: An Arduino UNO R3, the most common of the Arduino microcontroller boards.

The Hardware

Although there are now numerous variations on the Arduino (also known as "variants"), the most common Arduino, the UNO, consists of an Atmel ATmega328 8-bit microcontroller with a clock speed of 16 MHz. The ATmega328 has 32 KB of Flash memory, 2 KB of Static RAM (SRAM), and 1 KB of EEPROM onboard. For I/O, the Arduino has 14 digital I/O pins with 6 of these pins capable of Pulse Width Modulation (PWM), and six 10-bit Analog Inputs that can also be used for digital I/O pins. Two of the digital pins also directly support external hardware interrupts, and all 24 I/O pins support pin state change interrupts, allowing external hardware control of program execution.

Typically powered via the USB programming port, with its low current drain and onboard power regulator, the Arduino is ideally suited for battery powered projects. The Arduino supports multiple communication protocols, including standard Serial, SPI, I^2C, and One-Wire. Designed for expandability, the Arduino I/O and power connections are brought out to a series of headers on the main board. The header layout is standard among the majority of the UNO-type boards and many of the basic Arduino add-on boards, known as Shields, can be plugged directly into these headers and stacked one on top of the other, providing power and I/O directly to the Shield without any additional wiring needed.

Many types of shields are available, including all manner of displays, Ethernet, WiFi, Motor Driver, MP3, and a wide array of other devices. My personal favorite is the Prototyping shield, which allows you to build your own interface to an even wider array of Arduino-compatible components, modules, and breakout boards such as GPS, Real Time Clock, Compass, Text to Speech, and Lightning Detection modules along with an endless list of sensors such accelerometers, pressure, humidity, proximity, motion, vibration, temperature, wind speed, and many more.

History

As living proof that necessity is the mother of invention; the Arduino was created at the Interaction Design Institute Ivrea, in the northern Italian town of Ivrea. Originally designed as an inexpensive Open Source tool for students, replacing the more expensive and less powerful Parallax "Basic Stamp" development platform then used by students at the institute, the Arduino began as a thesis project in 2003 by artist and design student, Hernando Barragán, designed for a non-technical audience.

This project, known as Wiring, was designed on a ready-to-use circuit board with an Integrated Development Environment (IDE) based on the Processing Language created by Ben Fry and one of Barragán's thesis advisors, Casey Reas. Wiring was then adapted in 2005 by a team co-founded by another of Barragán's thesis advisors, Massimo Banzi. This team consisted of Hernando Barragán, Massimo Banzi, David Cuartielles, Dave Mellis, Gianluca Marino, and Nicholas Zambetti. Their goal was to further simplify the Wiring platform and design a simple, inexpensive Open Source prototyping platform to be used by non-technical artists, designers, and others in the creative field. Banzi's design philosophy regarding the Arduino is best outlined in his quote "Fifty years ago, to write software you needed people in white aprons who knew everything about vacuum tubes. Now, even my mom can program".

Unfortunately, at the same time, due a lack of funding the Institute was forced to close its doors. Fearing their projects would not survive or be misappropriated, the team decided to make the entire project Open Source. Released under the Open Source Creative Commons license, the Arduino became one of the first, if not the first, Open Source hardware products. Needing a name for the project, the team decided to name it Arduino; after a local pub named "Bar Di Re Arduino" which honors the memory of Italian King Arduin.

Everything about the Arduino is Open Source. The board designs and schematic files are Open Source, meaning that anyone can create their own version of the Arduino free of charge. The Creative Commons licensing agreement allows for unrestricted personal and commercial derivatives as long as the developer gives credit to Arduino and releases their work under the same license. Only the name Arduino is trademarked, which is why the various Arduino-compatible boards have names like Iduino, Ardweeny, Boarduino, Freeduino, etc. Typically these boards are fully compatible with their official Arduino counterpart and may include additional features not on the original Arduino board.

Massimo Banzi's statement about the Arduino project, "You don't need anyone's permission to make something great", and Arduino team member David Cuartielles quote, "The philosophy behind Arduino is that if you want to learn electronics, you should be able to learn as you go from day one, instead of starting by learning algebra" sums up what has made the Arduino so popular among hobbyists and builders. The collective knowledge base of Arduino sketches and program libraries is immense and constantly growing, allowing the average hobbyist to quickly and easily develop complex projects that once took mountains of datasheets and components to build. The Arduino phenomenon has sparked the establishment of a number of suppliers for add-on boards, modules, and sensors adapted for the Arduino. The current Arduino team consisting of Massimo Banzi, David Cuartielles, Tom

Igoe, Gianluca Martino, and David Mellis has continued to expand and promote the Arduino family of products.

Since its inception, the Arduino product line has been constantly expanding to include more powerful and faster platforms such as the recently released Arduino UNO R4 series with Renesas RA4M1 and ESP32-S3 microcontrollers, the ARM-Cortex M-series microcontroller-based MKR family, and the enhanced Arduino Nano family. Because of this ongoing expansion of the microcontrollers that support the Arduino "ecosystem", the Arduino and all of its variations now have the power needed to support processing-intensive applications and high speed communications.

What is Open Source?

Generally speaking, Open Source refers to software in which the source code is freely available to the general public for use and/or modification. Probably the best example of Open Source software is the Linux Operating System created by Linus Torvalds. Linux has evolved into a very powerful operating system, and the vast majority of applications that run on Linux are Open Source. A large percentage of the web servers on the Internet are Linux-based, running the Open Source Apache Web Server. The popular Firefox web browser is also Open Source, and the list goes on. Even the Android phone operating system is based on Linux and is itself Open Source. This ability to modify and adapt existing software is one of the cornerstones of the Open Source movement and is what had led to its popularity and success.

The Arduino team took the concept of Open Source to a whole new level. Everything about the Arduino, hardware and software, is released under the Creative Commons Open Source License. This means that not only is the Integrated Development Environment (IDE) software for the Arduino Open Source, the Arduino hardware itself is also Open Source. All of the board design files and schematics are Open Source, meaning that anyone can use these files to create their own Arduino board. In fact, everything on the Arduino website, www.arduino.cc, is released as Open Source. As the Arduino developer community grows, so does the number of Open Source Applications and add-on products, also released as Open Source. While it may be easier to buy an Arduino board, shield or module, in the vast majority of cases, everything you need to etch and build your own board is freely available for you to do as you wish. The only real restriction is that you have to give your work back to the Open Source Community under the same Open Source Licensing. What more could a hobbyist ask for? Everything about the Arduino is either free or low cost. You have a whole community of developers at your back, creating code and projects that you can use in your own projects, saving you weeks and months of development. As you will see in some of the projects in this book, it takes longer to wire and solder things together than it does to actually get it working. That is the true power of Open Source, everyone working together as a collective, freely sharing their work, so that others can join in on the fun.

Open Source Licensing and How it Works

There are several main variations on the Open Source Licensing Model, but all are intended to allow the general public to freely use, modify, and distribute their work. The most common Open Sources License models you will encounter include the Gnu General Public

License (GPL), Lesser GPL (LGPL), MIT, and the Creative Commons Licenses. As a general rule, for the average hobbyist, this means you are free to do as you wish. However, there will always be those of us that come up with that really cool project we can package up and sell to finance our next idea. It is important for that group to review and understand the various license models you may encounter in the Open Source world.

The GNU GPL

As with all Open Source Licensing models, the GNU General Public License (GPL) is intended to guarantee your freedom to share, modify, and distribute software freely. Developers who release software under the GPL desire their work to be free and remain free, no matter who changes or distribute the program. The GPL allows you to distribute and publish the software as long as you provide the copyright notice, disclaimer of warranty, and keep intact all notices that refer to the license. Any modified files must carry prominent notices stating that you changed the files and the date of any changes.

Any work that you distribute and publish must be licensed as a whole under the same license. You must also accompany the software with either a machine-readable copy of the source code or a written offer to provide a complete machine readable copy of the software. Recipients of your software will automatically be granted the same license to copy, distribute, and modify the software. One major restriction to the GPL is that it does not permit incorporating GPL software into proprietary programs.

The copyright usage in the GPL is commonly referred to as "copyleft", meaning that rather than using the copyright process to restrict users as with proprietary software, the GPL copyright is used to ensure that every user has the same freedoms as the creator of the software.

There are two major versions of the GPL, Version 2, and the more recent Version 3. There are no radical differences between the two versions, the changes are primarily to make the license easier for everyone to use and understand. Version 3 also addresses laws that prohibit bypassing Digital Rights Management (DRM). This is primarily for codecs and other software that deals with DRM content. Additional changes were made to protect your right to "tinker" and prevent hardware restrictions that don't allow modified GPL programs to run. In an effort to prevent this form of restriction, also known as Tivoization, Version 3 of the GPL has language that specifically prevents such restriction and restores your right to make changes to the software that works on the hardware it was originally intended to run on. Finally, Version 3 of the GPL also includes stronger protections against patent threats.

The Lesser Gnu General Public License (LGPL)

The LGPL is very similar to the GPL, except that it permits the usage of program libraries in proprietary programs. This form of licensing is generally to encourage more widespread usage of a program library in an effort for the library to become a de-facto standard, or as a substitute for a proprietary library. As with the GPL, you must make your library modifications available under the same licensing model, but you do not have to release your proprietary code. In most cases, it is preferable to use the standard GPL licensing model.

The MIT License

Originating at the Massachusetts Institute of Technology, the MIT license is a permissive free software license. This license permits reuse of the software within proprietary software, provided all copies of the software include the MIT license terms. The proprietary software will retain its proprietary nature even though it incorporates software licensed under the MIT license. This license is considered to be GPL-compatible, meaning that the GPL permits combination and redistribution with software that uses the MIT License. The MIT license also states more explicitly the rights granted to the end user, including the right to use, copy, modify, merge, publish, distribute, sublicense, and/or sell the software.

The Creative Commons License

There are multiple versions of the Creative Commons License, each with different terms and conditions:

1. Attribution (CC BY) - This license allows others to distribute, remix, tweak, and build upon a work, even commercially, as long as they credit the creator for the original creation.
2. Attribution-NonCommercial (CC BY-NC) – This license allows others to remix, tweak, and build upon a work non-commercially. While any new works must also acknowledge the creator and be non-commercial, any derivative works are not required to be licensed on the same terms.
3. Attribution-ShareAlike (CC BY-SA) – This is the most common form of the Creative Commons License. As with the Attribution license, it allows others to distribute, remix, tweak, and build upon a work, even commercially, as long as they credit the creator for the original creation and license their new creation under the same license terms. All new works based on yours convey the same license, so any derivatives will also allow commercial use.
4. Attribution-NonCommercial-ShareAlike (CC BY-NC-SA) – This license allows others to distribute, remix, tweak, and build upon a work non-commercially, as long as they credit the creator and license their new creations under the identical licensing terms.
5. Attribution-No Derivs (CC BY-ND) – This license allows for redistribution, both commercial and non-commercial, as long as it is passed along unchanged and in its entirety, with credit given to the original creator.
6. Attribution-NonCommercial-NoDerivs (CC BY-NC-ND) – This is the most restrictive of the Creative Commons licenses, only allowing others to download a work and share them with others as long as they credit the creator. Works released under this license cannot be modified in any way, nor can they be used commercially.

The Arduino is released under the Creative Commons Attribution-ShareAlike (CC BY-SA) license. You can freely use the original design files and content from the Arduino website, www.arduino.cc, both commercially and non-commercially, as long as credit is given to Arduino and any derivative work is released under the same licensing. So, if by chance you do create something that you would like to sell, you are free to do so, as long as you give the appropriate credit to Arduino and follow the requirements outlined in the FAQ on the Arduino website, as you may not be required to release your source code if you follow specific guidelines. If you include libraries in your work, be sure you use them within their licensing guidelines. The core Arduino libraries are released under the LGPL and the Java-based IDE is released under the GPL.

It is this Open Source licensing that has made the Arduino so popular among hobbyists. You have the freedom to do just about anything you want and there are many others developing code and libraries you can freely incorporate in your code, which helps make developing on the Arduino so much fun. For example, I know very little about Fast Fourier transforms, but there is a fully functional library out there just waiting for me to come up with a way to use it. That's the beauty of the Arduino and Open Source. You don't have to be a programming genius to create fully functional projects as long as you have the entire Open Source developer community at your back. And, when you do start creating wonderful new things, remember to share them back to the community, so that others following in your footsteps can benefit from your work and create wonderful new things of their own.

Chapter 2 • Arduino Boards and Variants

Choosing which Arduino board to use in your project can be a daunting task. Since the Arduino was first introduced in 2005, many new Arduino and Arduino variant boards have been developed, each with its own set of features and enhancements. Some of the older boards have gone by the wayside and been discontinued, but due to the Open Source nature of the Arduino, many of these older boards can still be found online. In this chapter, we will take a look at a few of the various Arduino boards that are currently available to help you determine which board would be best suited for your Arduino project ideas.

The Arduino UNO

When you talk about Arduino boards, the Arduino UNO (Figure 2.1) is often the first thing that comes to mind. The UNO is considered to be the original "mainstream" Arduino board and is the Arduino board that most people start out with. The majority of the existing libraries and sketches are designed to work with the UNO, so you have a wide variety of existing code to work with as you start out on your Arduino adventures. The Atmel ATmega328 series of processors has formed the core of the Arduino environment for a number of years, however, newer, faster, and more powerful processors are beginning to dominate the Arduino family. That being said, I still prefer to use the UNO in many of my projects, primarily for its low cost, simplicity, and ease of use.

Figure 2.1: The Arduino UNO R3.

The UNO uses an ATmega328 series microcontroller running at a clock speed of 16 MHz. It has 14 digital I/O pins and six 10-bit Analog-to-Digital pins that can also be used for digital I/O, if desired. Six of the digital I/O pins support pulse width modulation (PWM). Pulse Width Modulation allows you to control the pulse width of a square wave on a digital I/O pin to do things like dim an LED or generate an audio tone. Two of the digital I/O pins can be configured to support external interrupts for hardware program control, and all 20 I/O pins can be configured to provide a program interrupt when the I/O pin changes state. Interrupts can be pretty handy, and we'll talk more about them later.

The Arduino UNO has three types of onboard memory: Flash memory, Static Random Access memory (SRAM), and Electrically-Erasable Programmable Read-Only memory (EEPROM). Flash memory is rewritable memory that is primarily used to store your Arduino programs, known as sketches. Flash memory is semi-permanent and retains its contents even when power to the Arduino is turned off. You can rewrite the contents of Flash memory approximately 100,000 times, meaning that you can use the same Arduino board in dozens, and even hundreds of projects simply by uploading a different sketch. Flash memory can also be used to hold data that doesn't change, such as lookup tables, text, and other constants in order to save valuable SRAM space through the use of the PROGMEM keyword. SRAM is used to hold your program and system variables, and its contents are volatile, meaning that the contents are lost when the Arduino is reset or powered off. The onboard EEPROM can be used to retain data such as calibration values and similar settings between reset or power cycles. As with Flash memory, the Arduino EEPROM has a lifetime of approximately 100,000 write cycles. The Arduino UNO has 32 KB of Flash, 2 KB of SRAM, and 1KB of EEPROM onboard.

The Arduino UNO can be powered through the USB port or by 7-20 VDC on either its onboard DC power jack or connected to the Vin pin on the board itself. The ATmega-series microcontrollers support the industry-standard Serial Peripheral Interface (SPI), and Inter-Integrated (I^2C) bus communication protocols. The UNO, as with many other Arduino boards, has a standard 2.7 x 2.1 inch (58.6 mm x 53.3 mm) footprint. Female headers on the edges of the board allow for the stacking of add-on interface boards, known as shields, without the need for additional wiring.

The Arduino Nano

Figure 2.2: The Arduino Nano.

Not all boards come in the standard Arduino footprint that allows the use of shields. Some, such as the Arduino Nano (Figure 2.2), Mini (Figure 2.3), and Pico (Figure 2.4), are much smaller, yet have the same functionality and features of their larger brothers. These smaller boards allow you to build smaller, more compact projects, yet still have access to all of the standard libraries and code bases available for the UNO. These baby Arduinos, some not much larger than a postage stamp or two, are typically based on the ATmega328 and have the same features and functionality of the Arduino UNO.

Figure 2.3: The Mini Uno (photo courtesy David Mellis, via Wikimedia Commons).

Figure 2.4: The Arduino Pico.

The SparkFun Microview

Taking the concept of tiny Arduino to the next level, SparkFun has combined the ATmega328 with a 64 x 48 pixel Organic LED (OLED) display. The SparkFun Microview – OLED (Figure 2.5) is not much larger than a digital watch. Once programmed using the external USB programming adapter (Figure 2.6), all the Microview needs is 3.3 to 16 V DC power and you're off and running.

Figure 2.5: The SparkFun Microview.

Figure 2.6: The SparkFun Microview USB Adapter.

Wearable Arduinos

One of the newer applications for the Arduino is in the area of wearable electronics. While you may not immediately think that wearable electronics projects have a place in ham radio, their small size lends themselves to small projects, and things such as electronic name badges, and other interesting ham "wearables". Through the use of conductive thread, these wearable Arduinos can even be sewn into clothing and powered by small rechargeable Lithium-polymer (LiPo) batteries. While the Arduino Gemma (Figure 2.7) is based on the ATtiny85, most of the boards in this category, including the original Lilypad Arduino (Figure 2.8), are based on the ATmega328, with the major differences between them being size, available I/O pins, and type of power/battery and programming connections. These boards are programmed using the FTDI interface adapter or via an onboard micro-USB connector. The FTDI (Future Technology Devices International) adapter is a module that converts the USB connector from your workstation to a 6-pin DuPont-type header. A typical FTDI adapter is shown in Figure 2.9. The Lilypad Arduino SimpleSnap (Figure 2.10) is unique in that it has an onboard LiPo battery charging circuit, and is designed using conductive snap connectors, allowing you to remove the board from the project itself for washing or use in other projects.

Figure 2.7: The Arduino Gemma (photo courtesy Arduino.cc).

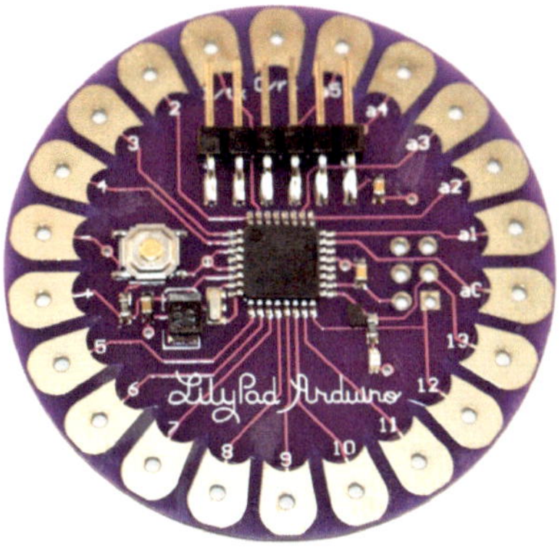

Figure 2.8: The Arduino Lilypad.

Figure 2.9: A typical FTDI adapter module.

Figure 2.10: The Arduino SimpleSnap Gemma (photo courtesy Arduino.cc).

The ATmega32u4 Series

Figure 2.11: The Arduino Leonardo.

Beginning with the Arduino Leonardo (Figure 2.11), the ATmega328 processor was replaced with the ATmega32u4. While nearly identical in functionality with the ATmega328, the ATmega32u4 eliminates the need for the UNO's ATmega16u2 external USB interface chip and allows the Leonardo to appear to a connected computer as a mouse and/or keyboard. The Leonardo and others in this class also feature 20 digital I/O pins instead of the UNO's 14, twelve analog input pins versus the six on the UNO, and seven of the digital I/O pins support PWM instead of the UNO's six. The ATmega32u4 also includes 2.5KB of SRAM, 512 bytes more than that of the ATmega328-series.

Figure 2.12: The Arduino Micro (photo courtesy Geek3, via Wikimedia Commons).

The Arduino Micro (Figure 2.12) brings the same functionality of the Leonardo to the smaller footprint of the Nano and Mini, while the Arduino Lilypad USB (Figure 2.13) is an ATmega32u4 version of the Lilypad often used in wearable Arduino projects.

Figure 2.13: The Arduino Lilypad USB (photo courtesy Aduino.cc).

The Arduino Mega Series

Figure 2.14: The Arduino Mega 2560.

The Arduino Mega series offers substantially more memory and I/O along with other features. Based on the 16 MHz ATmega2560 processor, the Mega 2560 (Figure 2.14) packs a punch with 54 digital I/O pins, sixteen 10-bit analog inputs, and four hardware serial ports on a larger 4 x 2.1-inch (101.6 mm x 53.3 mm) footprint. Six of the digital I/O pins can be configured for external hardware interrupts, and 15 of the digital I/O pins can provide PWM output. The I/O headers are compatible with most of the shields designed for the UNO and similar Arduinos, although care must be taken to ensure that the desired shield pin utilization matches the Mega I/O pin layout. The Mega2560 ups the Flash memory to 256 KB, the SRAM to 8 KB, and the EEPROM to 4KB. A smaller version of the Mega 2560 is now available, known as the Mega Mini (Figure 2.15). The Mega Mini is functionally identical to the Mega 2560 with a much smaller footprint.

Figure 2.15: The Mega Mini.

More Powerful Boards

Moving beyond the somewhat limited processor and memory capability of the ATmega328, newer Arduinos are incorporating more powerful processors, with more memory and other features. One of the nice things with these newer boards is that the Arduino IDE now has the capability of installing software "plug-ins" in order to support these newer processors. This means that you can create your sketch for one type of processor, and with a simple change in your IDE settings, you can compile the same sketch for another processor type without having to modify your sketch, other than to possibly take advantage of features inherent in the new board type. The first in this group of new Arduinos we will take a quick look at are the ones based on the ARM Cortex-M0+ processor. Care must be taken when working with these processors, as the ARM Cortex M0+ runs at 3.3 Volts and the I/O pins are generally not 5 Volt tolerant, meaning you can burn out your board if you apply 5 Volts on any of the I/O pins.

The Arduino Zero

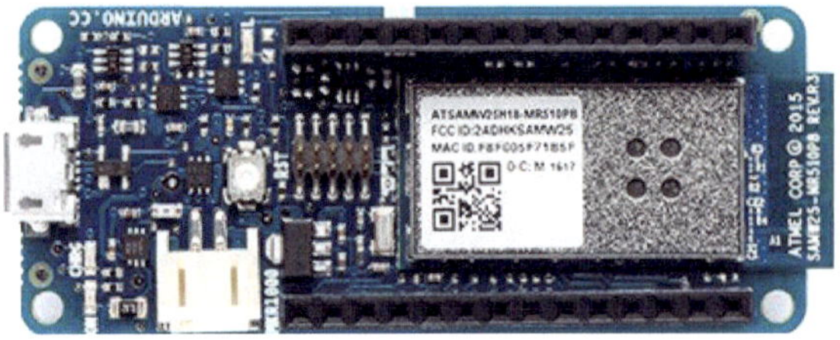

Figure 2.16: The Arduino Zero (photo courtesy Aduino.cc).

The Arduino Zero (Figure 2.16) uses the standard Arduino footprint and is powered by the SAMD21 32-bit ARM Cortex-M0+ processor. The Arduino Zero runs at 48 MHz and has 256 KB of Flash memory and 32 KB of SRAM. Like many of the newer Arduino processors, there is no onboard EEPROM. The Zero has 20 digital I/O pins, 12 of which support PWM, and six 12-bit analog input pins. The Zero also has one 10-bit Digital-to-Analog (D/A) pin and a hardware UART for serial communication. As with the older Arduino boards, the Zero also supports pin interrupts, along with the SPI and I²C communication protocols. The Arduino Zero also incorporates an embedded step-by-step debugger on board, allowing for in-depth program debugging.

The Arduino MKR1000 and MKRZero

Figure 2.17: The Arduino MKR1000 (photo courtesy Arduino.cc).

The MKR 1000 WiFi (Figure 2.17) and MKR Zero (Figure 2.18) provide the functionality of the Arduino Zero board in a smaller footprint. Both run at 48 MHz and have 256 KB of Flash and 32KB of SRAM. The MKR Zero has 22 digital I/O pins, twelve of which support PWM along with seven program-selectable 8 to 12-bit analog input pins. Both the MKR Zero and the MKR 1000 WiFi have a 10-bit D/A pin, one hardware UART serial pin, along with a JST connector and charging circuit for a 3.7 V LiPo battery. The MKR 1000 has 8 available digital I/O pins along with an integrated WINC5100 2.4 GHz IEEE 802.11 b/g/n WiFi module, while the MKR Zero has an onboard SD card slot.

Figure 2.18: The Arduino MKR Zero (photo courtesy Aduino.cc).

The PJRC Teensy 4.1

Figure 2.19: The Teensy 4.1.

Since the Arduino is Open Source, a number of vendors have been developing their own Arduino variants using these newer processors, including the ARM Cortex series. The Teensy series of Arduino compatible boards from PJRC is one of many newer, more powerful Arduino variants. Adafruit has created their Arduino-compatible Feather line, which we will cover separately in just a bit.

The Teensy 4.1 (Figure 2.19) board from PJRC is a small footprint board that uses an iMX-RT1062, containing a 32-bit ARM Cortex-M7 processor with a 32 and 64-bit floating point math unit (FPU) running at a whopping 600 MHz, with 8 MB of Flash and 1 MB of SRAM. These Teensy boards also include two USB ports, three CAN bus ports, two Inter-IC (I^2S) digital audio ports, and one S/PDIF digital audio port. It also has an SD memory card slot, along with three SPI and three I^2C bus interfaces. It also includes Cryptographic Acceleration, Random Number Generation, and a real-time clock onboard. There are also locations on the Teensy board to expand the Flash and SRAM memory with additional chips. The Teensy 4.1 increases the number of digital I/O pins to 55 and includes a 10/100 Ethernet interface onboard. One thing to be aware of with the Teensy 4.1 board is that not all the I/O pins are brought out to thru-hole solder connections. Some of the signals are brought out to small solder pads on the board, which does not lend itself very well to socket mounting.

The Espressif ESP32

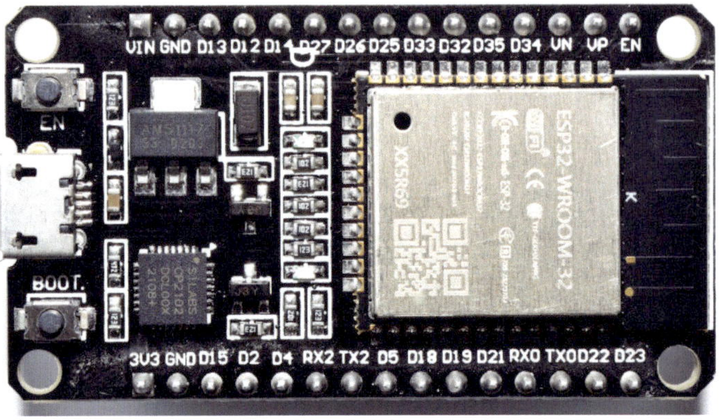

Figure 2.20: The Espressif ESP32.

The Espressif ESP32 (Figure 2.20) uses the Espressif ESP32 WiFi-enabled controller, which is a more powerful version of the Espressif ESP8266. The 32-bit ESP32 processor contains a dual core Tensilica LX6 microcontroller running up to 240 MHz. In addition to the onboard 2.4GHz IEEE 802.11 b/g/n WiFi controller, the ESP32 also comes with integrated dual-mode Bluetooth, supporting both classic and Bluetooth BLE devices. The ESP32 has 4MB of Flash and 520KB of SRAM, 28 digital I/O pins, with 16 pins supporting PWM, and 18 analog input pins. The ESP32 also has an onboard Hall-effect sensor, temperature sensor, and a microSD memory card slot, along with support for the SPI, I²C, and automotive CAN buses. The ESP32 is a 3.3 Volt device and the I/O pins are not 5 Volt tolerant.

The Adafruit Feathers

Figure 2.21: The Adafruit Feather 32U4.

Figure 2.22: The Adafruit Feather M0 Basic.

Adafruit has long been a leader in developing Arduino boards and modules. Adafruit's founder, Extra-Class Limor "Ladyada" Fried, AC2SN, passed all 3 of her amateur radio license tests in a single day. It therefore comes as no surprise that the Adafruit Feather series (Figure 2.21 and Figure 2.22) of Arduino-compatible boards support a wide array of WiFi, Bluetooth, and other methods of RF communications. The Adafruit Feathers are available with the Arduino 32u4 that is used in the Arduino Leonardo and other boards, the ARM Cortex-M0, M0+, and M3, and the ESP8266, all on a smaller Arduino Nano-style footprint. Depending on which version you're looking at, the Adafruit Feather supports Bluetooth LE,

2.4 GHz IEEE 802.11 b/g/n WiFi, and more recently, the license-free 433 MHz and 868/915 MHz Industrial/Scientific/Medical (ISM) bands, including 100mW Long Range (LoRa), capable of communication over distances up to 20 km. Since the memory and I/O specifications are different for each variant of the Feather, the best way to determine which Feather is right for you would be to visit the Adafruit website at www.adafruit.com. The Feather also has a wide array of add-on boards available. These boards, known as "FeatherWings", are similar to the shields used on the larger Arduinos. All the Adafruit Feathers are supported in the Arduino IDE simply by downloading and installing the Adafruit plug-ins for the Arduino IDE.

The STM32

Figure 2.23: The STM32.

The STM32 (Figure 2.23) is a family of 32-bit microcontrollers from STMicroelectronics and is based on the ARM Cortex-M CPU. While there are multiple versions of the STM32, the most common version is the "Blue Pill", featuring a 72 MHz ARM Cortex M-3 CPU.

The Blue Pill has 64 KB of Flash and 20 KB of SRAM memory, along with 32 digital I/O pins, 12 of which support Pulse Width Modulation. The Blue Pill also has 14 analog input pins, three Serial UART ports, two SPI buses and two I²C buses. The Blue Pill version is available from online suppliers such as AliExpress usually for under $3.00.

The Arduino R4

Figure 2.24: The Arduino R4 Minima.

The newly released Arduino R4 series is a pin-compatible replacement for the standard Arduino UNO R3 board. The I/O pins are 5 Volt tolerant, meaning that you can literally plug the UNO R4 in place of your existing UNO R3 board, with only one minor exception. While the I/O pins are 5 Volt tolerant, they can only supply 8ma per pin instead of the UNO R3's 40ma.

Available in two versions, the R4 Minima (Figure 2.24), and the R4 WiFi (Figure 2.25), The UNO R4 Minima has a 32-bit 48 MHz Renesas RA4M1 CPU, while the R4 WiFi includes the RA4M1 along with a 32-bit 240 MHz Espressif ESP32-S3 CPU. The RA4M1 CPU has 256 KB of Flash, 32 KB of SRAM, and 8 KB of EEPROM memory. The ESP32-S3 processor has 512 KB of SRAM and 384 KB of ROM. The R4 WiFi also includes onboard WiFi and Bluetooth support. Both versions feature a program-selectable 10 to 14-bit A/D (Analog to Digital) converter for the analog input pins, as well as a 12-bit D/A (Digital to Analog) converter output pin and on-chip Real-Time Clock. The Arduino R4 CPU also features an on-chip Operational Amplifier (OP AMP) and a Controller Area Network (CAN) bus for use in automotive applications.

Figure 2.25: The Arduino R4 WiFi.

New Nanos

Figure 2.26: The Arduino Nano 33BLE.

The Nano 33 BLE (Figure 2.26) is a pin and footprint-compatible substitute for the original Arduino Nano, however the I/O pins operate at 3.3 volts and are not 5 volt tolerant. The Nano 33 BLE features a 32-bit 64 MHz ARM Cortex M4 processor with 1 MB of Flash and 256 KB of SRAM. It has eight 12-bit A/D pins instead of the 10-bit A/D pins on the original Nano. The Nano 33 BLE also has a built-in 9-axis inertial measurement unit consisting of an accelerometer, gyroscope, and a magnetometer with 3-axis resolution, along with on-board Bluetooth and Bluetooth BLE support.

Figure 2.27: The Arduino Nano ESP32.

The Nano ESP32 (Figure 2.27) is also pin and footprint-compatible with the original Arduino Nano, however, as with the Nano 33 BLE, it operates on 3.3 volts and the pins are not 5 volt tolerant. The Nano ESP32 features a 32-bit 240 MHz Espressif ESP32 CPU with 128 MB of Flash, 512KB of SRAM, and 384KB of ROM. The Nano ESP32 also incorporates on-board WiFi and Bluetooth LE support.

The Raspberry Pi Pico/RP2040

Figure 2.28: The Raspberry Pi Pico.

The Raspberry Pi Pico (Figure 2.28) is Raspberry Pi's entry into the microcontroller arena. Based on the Raspberry Pi RP2040 processor, the Pi Pico is designed to be programmed using MicroPython, however, the programming language on the Pi Pico can be quickly changed and re-flashed with a simple press of the onboard BOOTSEL button, allowing the Pi Pico to run several other languages. These languages include CircuitPython, C/C++, several versions of BASIC, and includes official support for the Arduino programming environment.

The Pi Pico's RP2040 has a dual-core 133 MHz ARM Cortex-M0+ CPU, 2MB of Flash and 256KB of SRAM memory. It has 26 digital I/O pins, with 16 pins capable of Pulse Width Modulation along with three 12-bit analog inputs. It also has two SPI buses, two I²C buses, two serial UARTs, USB Host capability, and eight Programmable I/O (PIO) "state machines", along with an accurate on-chip real-time clock (RTC) and temperature sensor. The Raspberry Pi Pico runs on 3.3 volts and is not 5 volt tolerant. The Pi Pico W version also incorporates 2.4 GHz 802.11 b/g/n WiFi and Bluetooth support.

The Arduino Nano RP2040 Connect

Figure 2.29: The Arduino Nano RP2040 Connect.

The Arduino Nano RP2040 Connect (Figure 2.29) is an Arduino Nano footprint-compatible board based on the Raspberry Pi RP2040 CPU. Like the Pi Pico, the RP2040 Connect board runs on 3.3 volts and the I/O pins are not 5 volt tolerant.

The Nano RP2040 Connect features a 133 MHz dual-core RP2040 CPU with 16 MB of Flash and 264 KB of SRAM. It supports 802.11 b/g/n WiFi, Bluetooth, and Bluetooth BLE. Along with 22 digital I/O and 8 analog input pins, the RP2040 Connect also has a built-in microphone and 6-axis Inertial Measurement Unit (IMU), consisting of an accelerometer and gyroscope.

As with the Raspberry Pi Pico, the Arduino Nano RP2040 Connect also supports the Arduino and MicroPython programming environments.

With this wide array of Arduino boards to choose from, the sky is the limit for your Arduino projects. Newer, more powerful Arduino boards are being developed all the time, and with the smaller footprint and lower power requirements of some of these boards, your options for creating smaller, low-powered projects have been greatly expanded. Next, we'll discuss some of the add-on boards, modules, and components that would be of most interest to use in your own ham-related Arduino projects.

Chapter 3 • Arduino Shields, Modules and Devices

One of the biggest strengths of the Arduino lies in its ability to interface with a wide array of shields, modules, and devices. The standardized layout of the Arduino's onboard headers allows for add-on boards known as "Shields" to simply be plugged into the top of the Arduino board, providing instant access to displays, networking, and a number of other features that can be used with your Arduino projects. Through its use of the industry-standard Serial Peripheral Interface (SPI), Inter-Integrated Circuit (I^2C), and 1-Wire bus protocols, you can quickly and easily add a wide range of external modules and devices, including displays, Global Positioning System (GPS), Lightning Detection, Ethernet, and all manner of other sensors and devices.

The list of these external modules and devices is growing daily. To discuss all of the shields, modules, and devices for the Arduino that could be used in ham radio projects would easily take up an entire book by itself. Instead, this chapter will highlight the ones that I have personally found to have the most potential for use in ham-related projects. Also, rather than discuss every shield, module, and device for any one particular function, I have chosen a representative from each of the major functional roles I have encountered while building my projects. Please bear in mind that there are dozens of devices to choose from that function similarly. I encourage you to visit the websites of the various parts suppliers such as Adafruit, SparkFun, DFRobot, AliExpress, and Banggood (to name a few), and research the products they have to offer to determine which ones may be best suited for your project's needs.

Arduino Shields

Figure 3.1:The LCD Shield.

Shields are preassembled circuit boards you can plug into the headers that are on top of the Arduino board that allow instant access to all the features on the shield, without having to do any wiring whatsoever. You can even "stack" shields on top of each other, adding additional functionality with each additional shield. One can easily see how the shield stacking method can become rather unwieldy in a hurry and with each additional shield, the odds of an I/O pin usage conflict increases. You'll typically only want to stack one or two shields on top of the Arduino before deciding to use discrete modules and devices for your project.

Shields are most often used while you're becoming familiar with the Arduino and the project construction process. I often use shields in a prototyping or "proof of concept" role, rather than for finished projects.

Shields are typically supported by "libraries" and example code that allows you to develop functional sketches quickly and easily. An Arduino library is nothing more than a collection of files in a folder on your workstation that perform similarly to the drivers your PC needs to communicate with a new device. Libraries add functions and features for specific devices, allowing you to easily use the devices in your sketches without having to create the interface code on your own. Think of them as a "black box" between your sketch and the device. For example, you don't need to decipher the complex register settings and communication methods your sketch uses to tell the display controller chip device how to put information on a display, you just want it to show up on the right place on the display. Libraries do a lot of the heavy lifting for you, and are what help make the Arduino so easy to program and use.

In addition to being somewhat bulky and unwieldy, a major downside to shields is that they tend to be more expensive than the individual external modules and devices. If a shield fails, the cost to replace it would be significantly higher than replacing a simple module, so there is potentially a price to pay for the convenience of using a shield should something go wrong.

LCD Display Shield

The standard 16-character by 2-line LCD display shield (Figure 3.1) is one of the most commonly used displays when you're just starting out with the Arduino. This shield often includes a contrast adjustment potentiometer as shown in the upper left corner, and also features 6 pushbutton switches, allowing you to build simple Arduino projects quickly and easily. Note that the shield shown here is a low-cost generic shield, identifiable by the silk screen error on the "Right" pushbutton. This is one of the strengths of the Arduino's Open Source world. Many of the components used in Arduino projects are inexpensive and readily available from a multitude of suppliers thanks in part to the philosophy of the Arduino and Open Source hardware.

Color TFT Display Shield

Figure 3.2: The Touchscreen Color TFT and SD Memory Card Shield.

While there are a number of simpler display shields such as one for the 16-character by 2 line display, and another for the Nokia 5110-type LCD displays, I've found that the Color TFT displays are inexpensive and versatile as Arduino displays. The Color TFT display shield from Sainsmart (Figure 3.2) is actually a two-piece shield. The smaller shield board mounts on top of the Arduino and has a connector that allows you to select which size TFT display module you want to use with your project. The display module plugs into a connector on the shield itself. The Sainsmart display shield provides not only the TFT display capability, but also adds touchscreen support for the display module, and has an onboard SD memory card slot you can use to store images to display, thereby saving valuable Flash and SRAM memory on the Arduino.

Audeme MOVI™ Shield

Figure 3.3: The Audeme MOVI™ Shield.

The Audeme MOVI™ Shield (Figure 3.3) adds speech synthesis and speech recognition to your Arduino project. The MOVI shield allows you to use up to 150 separate full-sentence voice commands to control your project. It is programmed directly from the Arduino IDE and does not need an Internet connection or voice samples for training and is speaker independent.

The MOVI shield has a built-in microphone with automatic gain control that allows it to detect and recognize speech at a distance of up to 10 feet and the shield supports English, Spanish, and German.

Adafruit PowerBoost Shield

Figure 3.4: The Adafruit PowerBoost 500 Shield.

The Adafruit PowerBoost 500 Shield (Figure 3.4) allows you to use a 3.7 or 4.2 volt Lithium Ion or Lithium Polymer battery to power your Arduino project. It also includes an onboard 500 mA charging circuit that allows you to recharge the battery via the micro-USB port. The shield supports nearly any 35 mm x 62 mm x 5 mm capacity battery. The battery itself connects to the shield via a standard JST-type connector.

USB Host Shield

Figure 3.5: The USB Host Shield.

The USB Host shield allows you to interface USB devices such as a keyboard or mouse to your Arduino project. While some of the more recent Arduino boards include USB support on the board itself, the majority of the Arduino boards require an external shield or module similar to the one shown in Figure 3.5. As you start out with the Arduino, the USB Host shield is preferable to working with a USB Host module, simply because it's far easier to just plug a shield onto the top of your Arduino board than it is to wire up a module. As you continue to work with the Arduino, you may find that it's preferable to use the USB Host module since it allows for smaller, more compact projects that require USB support.

Ethernet Shield

Figure 3.6: The Ethernet Shield.

The Ethernet shield (Figure 3.6) allows you to link your Arduino projects to the Internet and create such projects as a Web Server or Internet of Things (IoT) device node using a wired 10/100 Ethernet port. As with the USB Host Shield, some of the more recent Arduino boards have onboard wired Ethernet capability. For the boards that don't have wired Ethernet, you can use either an Ethernet shield or an Ethernet module. As with the USB Host shield, you may find it preferable to use the Ethernet shield as you start out, then progress to the Ethernet module as you continue to build Arduino projects.

CAN-BUS Shield

Figure 3.7: The CAN-BUS Shield.

The CAN-BUS shield (Figure 3.7) allows you to interface to the industry standard automotive diagnostic bus found on most modern cars and some newer machine tools. The CAN-BUS shield allows you to access your vehicle's Electronic Control Units (ECUs) and read data such as throttle position, engine RPM, speed, and many other vehicle operating parameters.

The shields shown here are just a few of the most commonly used shields. As you become more familiar with building Arduino projects, you'll probably want to create your projects using modules instead. An extended list of Arduino shields can be found at www.shieldlist.org.

Arduino Modules

As you progress with building Arduino projects, you'll probably want to use modules instead of shields. Shields tend to be somewhat limiting in their functionality, are bulky, and often a bit on the expensive side. By using just a handful of wires, you can add the same functionality of a shield with a smaller, less expensive board or "module". Listed below are some of the Arduino modules most commonly used in ham radio Arduino projects.

The LCD display

Figure 3.8: A 16-character by 2-line Serial LCD Display.

Like the LCD Display Shield, the LCD display module is commonly used when you begin building Arduino projects. One of the big advantages to using the display is that it's simple to use and very easy to interface. The LCD Display module typically is available in two sizes. A 16-character by 2-line version, and a 16-character by 4-line version. The only operational difference between the two is the 2 extra lines of text and the larger physical size of the 16x4 LCD version. The 16x2 version is usually adequate for most Arduino projects that use this type of display module.

In addition to the size choices, there are two types of the LCD display modules, one with a parallel interface and another that interfaces to the Arduino using the I^2C bus. My personal preference is to use the I^2C bus version, as this module only uses the Arduino's analog pins A4 and A5 that are used by the I^2C bus, and it doesn't need to use any of the Arduino's digital I/O pins, whereas the parallel version of this module requires a minimum of 6 digital I/O pins for the 4-bit mode of operation, and 10 digital I/O pins for the faster 8-bit mode. The deciding factor between the two is usually whether or not a project will need the A4 and A5 pins used by the I^2C bus and if you can afford to tie up the larger number of digital I/O pins. More often than not, you'll find that you'd much rather reserve the digital I/O pins for your project and sacrifice a little bit of speed for the lower number of I/O pins required by the I^2C bus version.

The Nokia 5110 Display

Figure 3.9: The Nokia 5110 LCD Display.

Originally used in the older Nokia cell phones, the Nokia 5110 84x48 pixel graphic LCD shown in Figure 3.9 is another one of the more popular Arduino displays. Easy to interface to the Arduino using five digital I/O pins, the Nokia 5110 is small, easily readable, low-power, and capable of displaying both graphics and up to six lines of 14-character per line text. While technically the Nokia 5110 is an SPI-bus capable display, more often than not it is connected to the Arduino directly to digital I/O pins and the Arduino's SPI bus digital I/O pins are not used.

The Nokia 5110 is based on the Phillips PCD8544 LCD controller, and while specified for 3.3 Volts, I have used the Nokia 5110 with supply voltages from 2.7 to 5 volts, without 3.3 to 5 volt level shifters, with no damage to the display. That being said, I wouldn't advise taking any chances and recommend that you always use level shifters when interfacing to 3.3 Volt devices. Contrast settings can vary widely from display to display, however most libraries allow you to adjust the contrast level via software. The Nokia 5110 also has four LED backlights that can be controlled through the use of an additional digital I/O pin or a resistor from the backlight pin to ground.

More recently, multiple variants of the Nokia 5110 have become available, with some minor electrical differences between them. Figures 3.10 and 3.11 show three different versions of the Nokia 5110 display currently available. Note that the power pin is different on the one in the center and that the pin designations may not exactly match up with the ones in the project schematics. The display on the far left is the one that I use in my projects that utilize the Nokia display.

While most of the Nokia 5110 displays use a current-limiting for the backlight LEDs, the Adafruit version uses a transistor and requires a logic-high level to enable the backlight. When wiring up the Nokia 5110, always use the pin labels and not the pin numbers, as these designations differ between the various Nokia 5110 versions. When in doubt, always build a simple test project on your breadboard to determine the correct wiring for the version of the Nokia display that you plan to use.

Figure 3.10: Front view of the various Nokia display types.

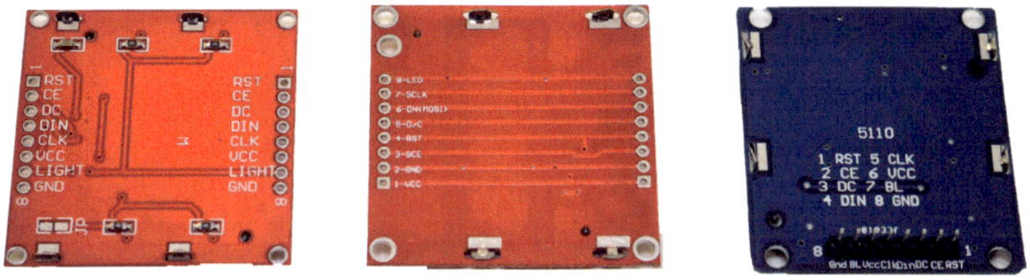

Figure 3.11: Back view of the various Nokia display types.

Organic LED (OLED) Displays

Figure 3.12: A 128 by 64 pixel Organic LED (OLED) module.

The organic LED (OLED) displays (Figure 3.12) are small graphic LED displays that communicate with the Arduino using either the SPI or I^{2}C bus. There are several versions of OLED displays available, primarily the 128x32 pixel and the 128x64 pixel versions. These are small displays, with the 128x32 version being about an inch wide by 3/8 inch high, and the 128x64 version is about an inch wide by 5/8 inch high. The 64x48 pixel SparkFun Micro OLED (Figure 3.13), is a mere 0.66 inches across and about 0.5 inches high.

An OLED display is comprised of tiny individual LEDs, and since it uses LED technology, does not need a backlight. OLED displays are very bright and can be used in sunlit applications. The libraries supporting the OLED displays allow you to display both text and graphics. The OLED display is an ideal choice when a small, low-power, bright, and clear sunlight-readable display is desired.

Figure 3.13: The SparkFun 64 by 48 pixel Micro OLED display module.

Thin Film Transistor (TFT) Color Displays

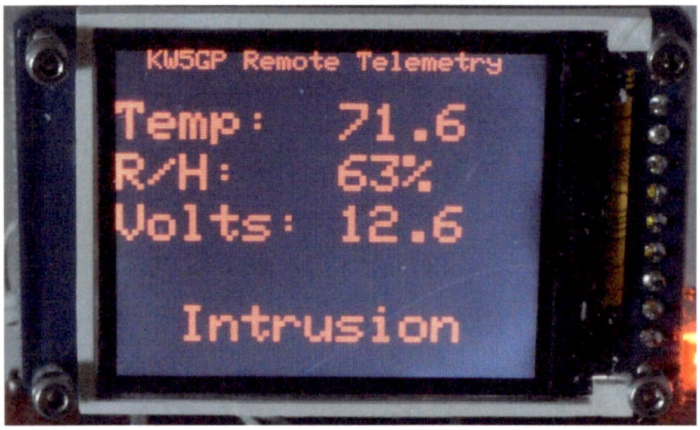

Figure 3.14: A 2.2-inch color TFT display module.

If you really want to spice up your Arduino display, a color TFT graphic display (Figure 3.14) is the way to go. TFT displays for the Arduino come in a variety of sizes, from 1.8 (128x160 pixels) inches (and even smaller) all the way up to a massive 7 inch (800x480) pixel display and offer up to 262,144 shades of color. The smaller 1.8, 2.2, and 2.8 inch versions of these displays communicate with the Arduino using either the SPI or I^2C bus. The 3.2 inch and larger displays will require an adapter shield due to the larger number of I/O pins to support these display sizes. Many of these display modules also include an onboard SD memory card slot, allowing you to save images on the module itself. Some of these displays also include a touchscreen, however, you will more than likely need a processor board such as the Arduino Mega to support both the display and touchscreen functions simultaneously.

While some of the TFT modules, such as the 1.8-inch ST7735-type TFT display from Adafruit, support 5 volts, the majority of TFT displays operate on 3.3 volts and may or may not be 5 volt tolerant on the I/O pins. To prevent damage to these displays, you should always ensure that you do not apply more than 3.3 volts to the module's power and I/O pins. A simple solution to this issue is to use a level shifter module, similar to the one shown in Figure 3.15.

Level Shifter

Figure 3.15: A generic TXB0108-type 8-channel Level Shifter.

While you can use a resistor divider to interface the digital I/O pins to a 3.3 volt device, it's usually easier and more reliable to use a level shifter module similar to the TXB0108-type shown in Figure 3.15. Also, a resistor divider only works in one direction whereas a level shifter module is bidirectional. Commonly available in a 4 or 8-channel configuration, the level shifter module allows for simple interfacing of 3.3 volt devices to a 5 volt Arduino, or vice versa.

Bluetooth Module

Figure 3.16: An HC-05-type Bluetooth Module.

For simple Bluetooth projects, you'll want to use the HC-05-type Bluetooth module (Figure 3.16). This module supports both the master and slave Bluetooth roles. In slave mode, you can pair your Arduino to a device such as a workstation or cell phone. In master mode you can pair and connect multiple Bluetooth slave modules to your Arduino.

Some of the currently available Bluetooth modules can be switched between master and slave mode from with your Arduino sketch. My personal choice is the ZS-040 version. The Arduino communicates with the Bluetooth module using standard TTL serial communications, usually at a default speed of 9600 baud.

Be careful and verify the specifications for your particular Bluetooth module, as some accept power in the range of 3.6 to 6 volts, but only allow a maximum of 3.3 volts on the data pins. You can use a simple resistor divider network, or a level shifter module to protect the data pins from overvoltage.

GPS Module

Figure 3.17: A typical GPS module.

With a GPS module such as the one shown in Figure 3.17, you can add all the power of a standard GPS (Global Positioning System) to your Arduino, providing such information as highly accurate time, latitude, longitude, altitude, speed, course, and other features of a standard GPS system. These GPS modules communicate with the Arduino via a standard TTL serial I/O port, typically with a default speed of 9600 baud, and they output standard National Marine Electronics Association (NMEA) NMEA-0183 messages. The Arduino GPS libraries parse the NMEA messages sent by the GPS into data that can be used in your Arduino sketches through the use of simple function calls.

Real-Time Clock Module

3.18: The DS3231 Real-Time Clock Module.

The DS3231 (Figure 3.18) is an inexpensive but extremely accurate real-time clock (RTC) module for the Arduino. A vast improvement on earlier real-time clock modules, the DS3231 chip has an internal temperature-compensated crystal oscillator (TCXO) and crystal. Since the Arduino does not have a real-time clock onboard, the DS3231 is ideal for those projects where you need to keep track of date and time, especially across power off and reset events. The DS3231 module communicates with the Arduino via the I^2C bus and has an onboard coin-cell battery to maintain the clock when power is off. As with earlier RTC modules, the DS3231 can keep track of time, day, month, and year, with built-in leap-year compensation up to the year 2100. The DS3231 also has a built-in temperature sensor with an accuracy of ± 3° C. As with most Arduino modules, the DS3231 is supported by many excellent libraries.

DS18B20 Temperature Sensor

Figure 3.19: The DS18B20 Temperature Sensor.

The Dallas Semiconductor (now Analog Devices) DS18B20 (Figure 3.19) is a programmable 9 to 12-bit resolution digital thermometer module that communicates via the 1-Wire Interface using a single wire for both power and data. Every DS18B20 module has a unique 1-Wire address, so multiple modules can function using just a single Arduino I/O pin.

The DS18B20 can measure temperature from -55°C to 125°C (-67°F to 257°F) with an accuracy of +/- 0.5°C (0.9°F) and also features an alarm function with non-volatile user-programmable upper and lower trigger points.

DHT20 Relative Humidity and Temperature Sensor Module

Figure 3.20: The DHT20 Relative Humidity and Temperature Sensor Module.

The DHT20 Relative Humidity and Temperature Sensor Module (Figure 3.20) is similar in function to the older discontinued DHT11/DHT22 sensors but uses the I^2C bus for communication. Based on the AHT20 temperature and relative humidity sensor, the DHT20 has a temperature sensing range of –40 °C to 80 °C (–40°F to 176°F) with an accuracy of ± 0.3°C (0.5 °F) and a relative humidity sensing range of 0% to 100% RH with a ±2% accuracy.

GY-906 Infrared Temperature Sensor

Figure 3.21: GY-906 Infrared Temperature Sensor Module.

The GY-906 (Figure 3.21) Infrared Temperature Sensor module is a non-contact sensor based on the MLX90614 sensor that allows temperature measurement between –70°C to 380°C or –94 to 716°F, with a resolution of 0.5 °C (0.9 °F) and a sensing distance of approximately 2 feet (60 cms).

BMP280 Air Pressure and Temperature Sensor

Figure 3.22: BMP280 Air Pressure and Temperature Sensor Module.

The BMP280 Air Pressure and Temperature Sensor (Figure 3.22) is a precision Bosch sensor that not only performs routine environmental measurements, but also doubles as an altitude and elevation sensor.

BME280 Air Pressure and Temperature Sensor

Figure 3.23: BME280 Air Pressure, Temperature, and Humidity Sensor Module.

The BME280 Air Pressure and Temperature Sensor (Figure 3.23) is similar to the BMP280 described above, however the BME280 module adds the ability to measure humidity in addition to air pressure and temperature.

Rain and Water Sensor

Figure 3.24: The Rain and Water Sensor Module.

The Rain and Water sensor (Figure 3.24) allows you detect the presence of rainfall or water. When water falls on the conductive plates of the sensor and by measuring the change in the resistivity of the conductive plates, the intensity of the rainfall can be determined. This sensor could also be used as a simple detector for the presence of water.

Wind Speed Sensor

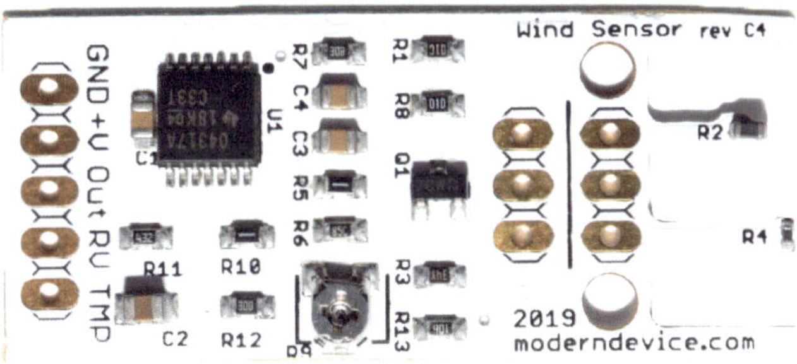

Figure 3.25: Wind Speed Sensor Module.

The Wind Speed Sensor (Figure 3.25) from Modern Device (www.moderndevice.com) is a low-cost anemometer without any moving parts. Using a technology known as "Hot Wire" wind speed measurement, this sensor heats an element to a constant temperature and then measures the electrical power required to maintain that temperature. This measurement technique is great for measuring low to medium wind speed. This sensor is quite sensitive, being able to detect a small puff of air at a distance of up to 2 feet. The "C" version shown here can accurately measure wind speed from 0 to 60 mph. The "P" version is even more accurate and can measure hurricane force winds up to 150 mph.

Lightning Detector Module

Figure 3.26: Lightning Detector Module.

Based on the AS3935 Franklin Lightning Detector chip from Austria Microsystems (now Sciosense), the Lightning Detector module (Figure 3.26) uses a proprietary on-chip signal processing algorithm to detect lightning at a distance up to 40 km (24.8 miles). It does this by analyzing the 500 kHz component of the lightning and looking for a specific signal pattern, or "signature", indicative of lightning. The AS3935 is capable of detecting both cloud-to-ground and cloud-to-cloud lightning. The AS3935 includes an embedded algorithm to reject man-made electrical noise and has software-selectable threshold settings. The AS3935 can statistically calculate the distance to the leading edge of the thunderstorm, along with the estimated strength of the lightning strike. The AS3935 communicates with the Arduino via either the SPI or I²C bus.

Ethernet Module

Figure 3.27: The ENC28J60 Ethernet Module.

The ENC28J60 Ethernet Module (Figure 3.27) allows you to quickly add wired 10/100 Mb/s Ethernet networking capability to your Arduino projects. Easily interfaced to the Arduino via the SPI bus, the Ethernet module is supported by a number of Arduino libraries and allows you to turn your Arduino into a web server, FTP file server, and even a Network Time Protocol (NTP) server.

Direct Digital Frequency Synthesis (DDS) Module

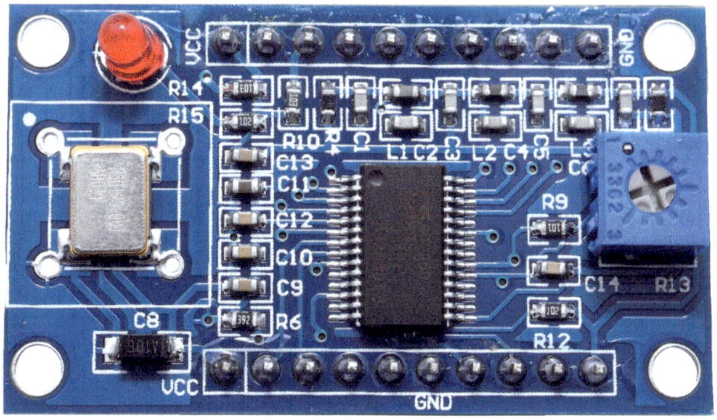

Figure 3.28: The AD9850 Direct Digital Frequency Synthesis (DDS) Module.

Prior to the development of the direct digital frequency synthesis module, about the only way to generate an RF frequency signal was to use a crystal or an unstable (relatively speaking) Resistive-Capacitive (RC) oscillator. Using a DDS module such as the AD9850 DDS module shown in Figure 3.28, you can programmatically generate highly stable waveforms up to 62.5 MHz with an accuracy of 0.0291 Hz. Other DDS modules support higher frequencies than the AD9850, however I have found the inexpensive AD9850 to be more than adequate for most Arduino projects. A DDS generates a smooth sine or square wave, while some DDS modules, such as the AD9833 (Figure 3.29) also allow you to generate triangle waves.

One of the major advantages of using a DDS module over processor-generated waveforms is that the DDS chip offloads all of the frequency and waveform generation from the processor and, in the case of the AD9850, uses a 10-bit digital-to-analog converter on the chip to produce clean, stable sinusoidal waveforms across the entire operating range. The AD9850 DDS also has the capability of shifting the phase of the output waveform from 0-2π (0-720°) and interfaces to the Arduino using the SPI bus. There are a number of libraries supporting the DDS modules, including a very simple and easy to use library for the AD9850 written by Paul Darlington, M0XPD.

Figure 3.29: The AD9833 Direct Digital Frequency Synthesis (DDS) Module.

Si5351 High Frequency Programmable Clock Generator Module

Figure 3.30: Si5351 High Frequency Programmable Clock Generator Module.

The Si5351 is similar to a DDS in that both generate highly precise frequencies. In the case of the Si5351, it can simultaneously generate up to eight different frequencies from 8 kHz to 160 MHz, although many of the inexpensive Si5351 modules currently available use a three output version of the chip. The generic Si5351 module shown in Figure 3.30 communicates with the Arduino using the I^2C bus and has three independent outputs, each capable of providing 3 volt peak-to-peak square waves. The Si5351 is a 3.3 volt device, so you will need level shifters on the Si5351 I/O pins if you are using a 5 volt Arduino.

One major difference between the AD9850 DDS module and the Si5351 is that the Si5351 only has a square wave output, as compared to the sine and square wave output of the AD9850 DDS. This means that an unfiltered RF signal generated by the Si5351 will also generate unwanted harmonics at multiples of the original frequency. For this reason, if you are going to use the Si5351 as part of a transmitter or VFO, for example, you will want to add a filtering section to smooth the square into a sine wave to reduce harmonic output. Adafruit and others have excellent libraries that support the Si5351 and allow you access to all of the features of the Si5351 chip.

DFRobot Gravity: Speech Synthesis V2.0 Module

Figure 3.31: The DFRobot Gravity: Speech Synthesis V2.0 Module.

The DFRobot Gravity: Speech Synthesis V2.0 Module (Figure 3.31) allows you to add natural sounding speech to your Arduino projects. Capable of speaking in English and Chinese, the Gravity communicates with the Arduino via I^2C or by using a standard TTL serial I/O port and includes an onboard audio amplifier and speaker. Supported by the DFRobot Speech Synthesis Library, you can control the volume, speed, voice pitch and inflection, and even sound modes to make the voice sound nearby, distant, robotic, underwater, and other choices.

DFRobot Gravity: Offline Language Learning Voice Recognition Module

Figure 3.32: DFRobot Gravity: Offline Language Learning Voice Recognition Module.

Adding voice recognition to your Arduino projects can take hands-free user interaction to a whole new level. For example, with a voice recognition module similar to the DFRobot Gravity: Offline Language Learning Voice Recognition Module (Figure 3.32), you can create a completely voice-controlled CAT (Computer Aided Transceiver) system for your shack. This module comes with 121 built-in fixed command words and supports the addition of 17 custom command words. The module is completely standalone and does not need an Internet connection to function.

The scary part about using this particular module, especially for me, is that the DFRobot web site's description includes a discussion on how any sound could be used as a custom command word, including cat meows. So now you can truly create cat-controlled Arduino CAT ham radio projects. I'm not really sure I want to give my cats that kind of control, as it would be like giving them opposable thumbs, nothing would be safe anymore. Scary as that may sound, still, the idea of using sounds as command words does have interesting possibilities for creating cool Arduino projects.

The DFRobot Voice Recognition module supports both I^2C and serial TTL communication and is also both 3.3 and 5 volt compatible.

MP3 Player Module

Figure 3.33: A generic MP3 Player Module.

For applications where you don't necessarily need a more expensive text-to-speech synthesizer, you can use an MP3 player module similar to the generic MP3-TF-16P Player shown in Figure 3.33 to play pre-recorded MP3 audio files. This module includes an onboard SD memory card slot used to store up to 32 GB worth of MP3 audio files that you can play back at a program-selectable sampling rate from 8 kHz up to 48 kHz. This module allows you to have up to 100 folders and 255 MP3 files with 30 levels of adjustable volume.

ISD1820 Voice Recorder and Playback module

Figure 3.34: The ISD1820 Voice Recorder and Playback module.

The ISD1820 (Figure 3.34) is a Voice Recorder and Playback module capable of recording and playing back a single 8 to 20 second message. This module can be operated standalone or interfaced to the Arduino via the SPI bus. The recording duration and sampling are controlled by a resistor connected to the ROSC pin on the ISD1820 chip and ground. Most ISD1820 modules have the recording and playback duration configured with a 100-kΩ resistor, for a recording duration of 10 seconds, a sampling rate of 6.4 kHz and a bandwidth of 2.6 kHz.

The ISD1820 module includes an onboard microphone and an on-chip 8 Ω / 0.5 W speaker driver, along with pushbuttons to active the recording and playback functions.

DFRobot Gesture Sensor Module

Figure 3.35: The DFRobot Gesture Sensor module.

The DFRobot GR10-30 Gesture Sensor is capable of accurately recognizing 12 hand gestures at a distance of up to 30 cm (11.8 inches) and has two interrupt pins to indicate if a gesture trigger occurs and if an object enters the recognition range. This module supports standard serial UART and I²C communication.

TI ADS1115 16-bit Analog to Digital (A/D) Converter Module

Figure 3.36: The TI ADS1115 16-bit A/D module.

The analog-to-digital (A/D) converter used for the analog input pins on the Arduino UNO and Nano is a single-ended 10-bit A/D, providing a count value of 0 to 1023 over a 5 volt range with respect to ground. For many of your Arduino projects, this resolution is more than adequate; however, there will be occasions where you may prefer to have higher resolution in your analog voltage measurements than the A/D onboard the Arduino allows. In the case of an antenna rotator position sensor, for example, you will want to read a position in degrees from 0 to 360° (or 450° in the case of some rotators). Using a 10-bit A/D, this leaves you with a resolution of about 1°. A 12-bit A/D module such as the Texas Instruments ADS1015 would provide a resolution of about 1/10°, and a 16-bit A/D module such as the Texas Instruments ADS1115 (Figure 3.36) would provide you with a resolution of about 1/100°. As you can see, the more bits your A/D has, the higher the resolution.

The TI ADS1015 12-bit A/D module can provide a count of 0 to 4095 on four single-ended, or two differential inputs, with a program-selectable sampling speed up to 3300 samples per second. The TI ADS1115 16-bit A/D module can provide a count range of 0-65535 on four single-ended, or two differential inputs, with a program-selectable sampling speed up to 860 samples per second. Both communicate with the Arduino using the I^2C bus, and feature six program-selectable gain settings, from a full-scale reading of 6.144 volts (warning, do not ever exceed VDD + 0.3 Volts on any input pin, so you will never be able to read more than the supply voltage at the 2/3 gain setting) all the way down to a full-scale reading of 0.256 volts at the 16x gain setting. There are a number of Arduino libraries supporting these modules that allow access and control of all the internal parameters of these modules and make using these modules in your Arduino projects a breeze.

MCP4725 12-bit Digital to Analog (A/D) Converter Module

Figure 3.37: The Microchip Technology MCP4725 12-bit D/A module.

Performing the reverse function of an A/D module, a digital-to-analog module (D/A) converts a digital value into an analog voltage representation. The one thing the Arduino UNO and Nano are missing is an onboard D/A converter, although some of the newer variants do have this feature built in. In trying to add this functionality to my Arduino projects, I have constructed several types of D/A converters, from a simple 8-bit resistive ladder all the way up to using an I^2C D/A module. The results of these tests showed that the Microchip MCP4725 12-bit I^2C D/A converter shown in Figure 3.37 to be my D/A of choice, both for its simplicity and ease of use.

The MCP4735 supports three I^2C bus speeds, 100 kb/s, 400 kb/s, and 3.4 MB/s and has eight user-selectable I^2C bus addresses. The MCP4725 also features an on-chip EEPROM that can be used to save and restore the chip's registers and configuration bit values between power off/on and reset events.

Using the MCP4725 Arduino library and sample sketches, you can quickly have your Arduino outputting analog voltages, sine waves, and even triangle waves, among other waveforms.

9 Degree of Freedom (9DOF) Sensor Module

Figure 3.38: A generic 9-Degree of Freedom Sensor module.

The "9-Degree of Freedom Sensor" is also known as a 9-axis Motion Tracking Device or IMU (Inertial Measurement Unit) sensor. These modules house a 3-axis gyroscope, a 3-axis accelerometer, and a 3-axis magnetometer on a single chip used to measure orientation, velocity, and gravitational force, allowing you to precisely track position and motion.

ACS712 Hall-Effect Current Sensor Module

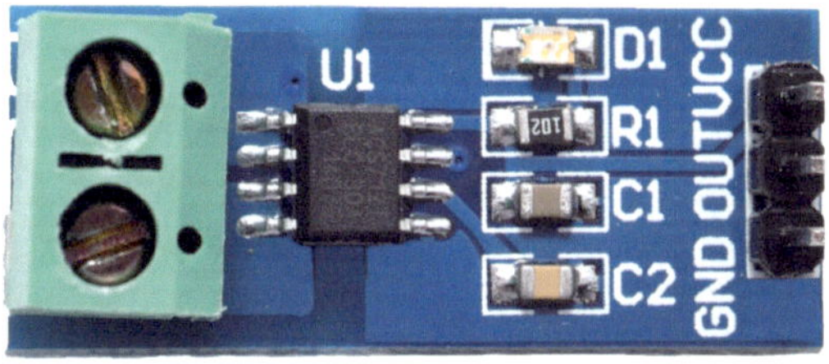

Figure 3.39: An ACS712 Hall-Effect 30A Sensor module.

Measuring current is something it seems that just about every ham does at some point or another. Fortunately, there are several choices available for measuring current from just a few milliamps all the way up to 50 amps and higher. The ACS712-based Current sensor module shown in Figure 3.39 can be used to measure DC or AC current up to 30A and outputs a voltage proportional to the current for connection to an Arduino's analog input pin.

The ACS712 chip uses the Hall Effect to detect the strength of a magnetic field in a conductor and provides an isolated output voltage proportional to the strength of the magnetic field induced by the current flowing through the circuit. This provides a safe method to measure current in your Arduino projects without subjecting the Arduino analog input pins to a voltage that could damage the Arduino. The ACS712 chip itself features 2.1 kV RMS of isolation, meaning that you can use it to measure the current in a high voltage (i.e., tube amplifier) circuit.

Non-Invasive AC Current Sensor

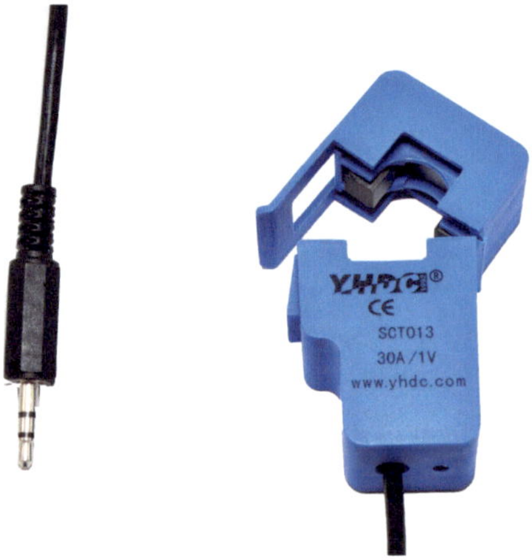

Figure 3.40: A typical non-invasive AC current sensor.

The Non-Invasive AC Current Sensor uses a split-core transformer to sense the strength of the magnetic field of a wire. The alternating current in the AC circuit creates an alternating magnetic field that then induces a corresponding voltage in the split core transformer that can be measured using an analog input pin on the Arduino.

The main benefit of using this type of current sensor is that it does not need to be inserted into the circuit under test and can be easily added or removed.

Ferroelectric RAM (FRAM) Memory Module

Figure 3.41: The Adafruit FRAM Memory Module.

The Ferroelectric RAM memory module shown in Figure 3.41 has 32 KB of storage and can communicate with the Arduino over the I²C bus at speeds up to 1 MHz. Ferroelectric RAM is a form of non-volatile random-access memory. Similar to the ferrite core memory technology used in the early days of computers, FRAM is capable of high-speed reading and writing, but unlike Static RAM (SRAM) or Flash memory, FRAM can retain data for 95 years.

Unlike Flash memory that has a write limit of approximately 100,000 times, FRAM can be read/written 10 trillion times, giving you the best of SRAM and Flash memory in a single module.

SC16IS750 UART Module

Figure 3.42: The SC16IS750 SPI UART Module.

The SC16IS750 UART Module (Figure 3.42) is used to interface to the Arduino using either the SPI or I²C bus and provide serial UART (Universal Asynchronous Receiver Transmitter) functionality. This will allow you to communicate with TTL serial and, through the use of a level converter such as the MAX232 or MAX485, standard RS-232 and RS-485 devices.

Since the Arduino UNO's TTL serial communication pins are shared with the USB serial connection to a workstation, if you need to interface a serial device such as a GPS, you would either need to use the Software Serial library to emulate a hardware serial port or add an Arduino board or module that supports additional serial ports.

The SC16IS750 is fully compatible with the industry standard 16C450 UART which itself goes all the way back to the original 8250 series of UARTs. The 16IS750 module supports data rates up to 5 Mb/s and provides 8 additional programmable I/O pins.

The SC16IS750 features a 64-character buffer on both the transmit and receive side and a programmable baud rate using a 14.7456 MHz crystal. It can also be configured for a word length between 5 and 8 bits, with the usual stop and parity bit selection.

In addition to the standard 8-bit serial communication mode for devices such as GPS modules, the programmable baud rate and the ability to communicate using 5-bit data words can be of great interest to hams. Radio Teletype (RTTY) typically uses 45.45 baud and a 5-bit data word (also known as Baudot code). Using the 16IS750 module, you can easily build an Arduino project that can send and receive RTTY. There's even a RTTY library for the Arduino to convert between Baudot and standard ASCII.

USB Host Module

Figure 3.43: A USB Host Module.

Technically, the USB Host module (Figure 3.43) is actually a USB host shield designed for the Arduino Pro-Mini. However, it doesn't use the standard Arduino UNO shield footprint and it's small enough to fit in many Arduino projects when used as a module.

Supported by the Arduino USB Host library, when this board is used as a module, you can easily provide USB host capability to your Arduino projects, allowing you to add standard USB HID (Human Interface Devices) devices such as keyboards, mice, etc. to your Arduino projects.

4x4 Membrane Keypad

Figure 3.45: The standard Arduino 4x4 Membrane Keypad.

The keypad most often associated with Arduino is the 4x4 (16 key) Membrane keypad shown in Figure 3.45. This keypad is made of a thin flexible membrane and connects to the Arduino using 8 digital I/O pins, 4 for the row sensing, and 4 for the column sensing. Because this keypad requires so many digital I/O pins, it is generally only used in applications that require a weatherproof, water-resistant keypad.

4x4 Analog Keypad

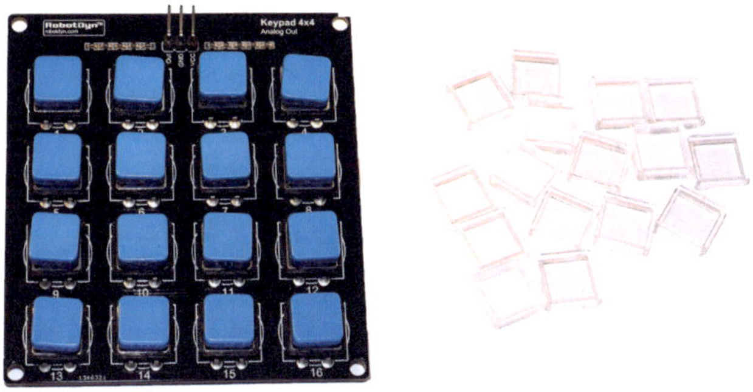

Figure 3.46: The standard Arduino 4x4 Analog Keypad.

Rather than tying up 8 digital I/O pins as in the case of the 4x4 Membrane keypad, the 4x4 Analog Keypad (Figure 3.46) from AliExpress uses a single analog pin to send the key data to the Arduino. This keypad uses a series of precision resistors in a 4x4 row and column arrangement to provide a unique voltage for each key pressed on the analog pin.

This keypad also features removable key caps, allowing you to create your own label for each individual key.

Adafruit 4x4 Elastomer Keypad, Monochrome Driver, and Enclosure

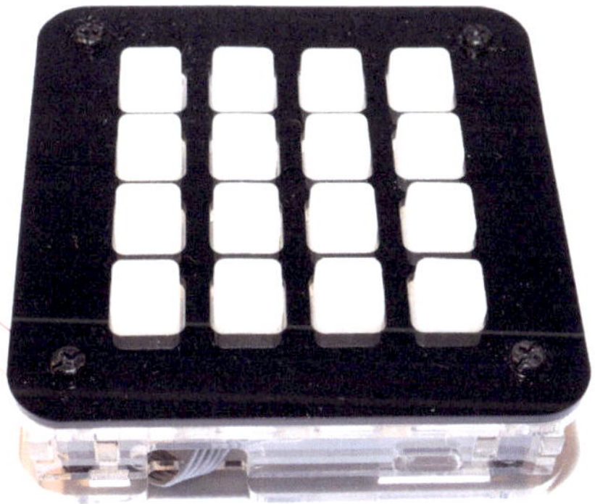

Figure 3.47: Adafruit 4x4 Elastomer Keypad.

The Adafruit 4x4 Elastomer Keypad (Figure 3.47) is a soft silicone elastomer 4x4 keypad that features a 4x4 matrix. Designed to be paired with their Adafruit Trellis Monochrome driver board that adds key lighting on a per-key basis, the Trellis board communicates with the Arduino via the I^2C bus.

When combined with the Adafruit 4x4 Trellis Feather Acrylic Enclosure, this keypad is a great choice for projects requiring a 4x4 keypad.

Addressable RGB LEDs

Figure 3.48: An Adafruit NeoPixel Addressable LED Ring.

Addressable RGB LEDs are available in all sizes of rings, ring segments, sticks, and strips. Figure 3.48 shows an RGB NeoPixel LED ring from Adafruit. Each LED pixel (Figure 3.49) typically consists of a control chip and a 5050-type RGB LED, with 256 levels of brightness and 24-bits of color control that can display 16.8 million different colors.

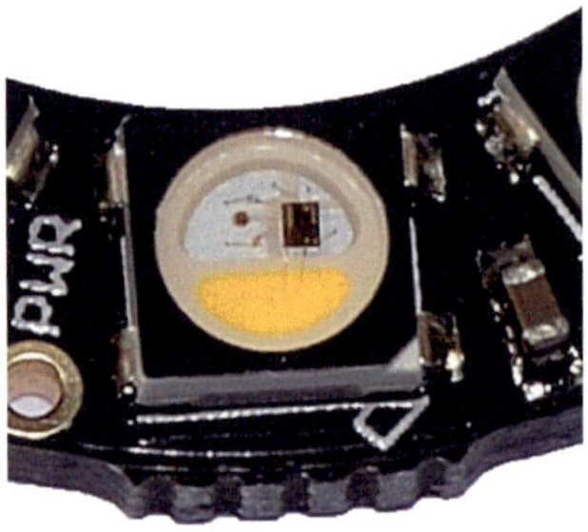

Figure 3.49: A close-up view of an RGB LED pixel showing the control chip and the LED .

eInk/ePaper Displays

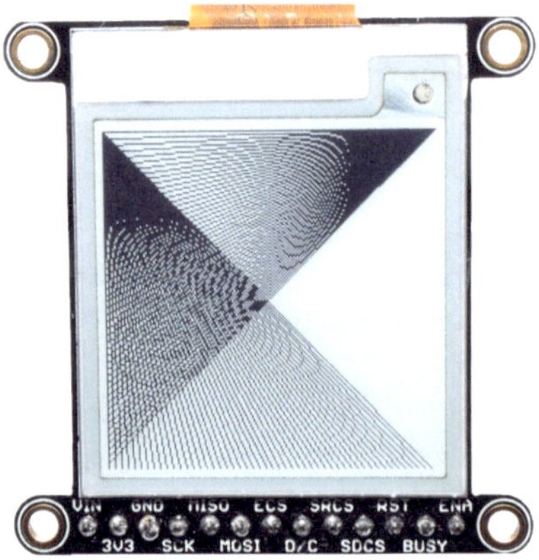

Figure 3.50: An Adafruit e-Ink display.

eInk/ePaper display modules similar to the one shown in Figure 3.50 are the latest thing in Arduino displays. Available in many sizes in both monochrome and color, these displays operate on very low power and even retain the displayed image when power is turned off.

eInk/ePaper display modules are comfortable to read and have a wider viewing angle than most display modules. They can even be read in direct sunlight without the image appearing to fade. eInk/ePaper displays work by rearranging charged pigment particles with an applied electric field.

While eInk/ePaper displays look great and have a number of advantages over other types of displays, it's important to note that there is a recommended limit on how often you can update or refresh the display. The current recommended upper limit is every 180 seconds (3 minutes). Exceeding this limitation could cause permanent damage to the eInk display itself.

While this list of Arduino shields and devices is lengthy, it's nowhere near a complete list of what you can add to your Arduino projects. I recommend checking the websites of suppliers such as Adafruit, SparkFun, DFRobot, and others on a regular basis to stay on top of what new products may be available to use in your Arduino projects.

Enclosures

Figure 3.51: An Arduino mounted in an Altoids™ mint tin.

Because of its small size, nearly anything can be used as an enclosure for your Arduino projects, such as the Altoids™ mint tin shown in figure 3.51. A quick search on eBay, Adafruit, SparkFun, and other suppliers offers a wide variety of enclosures designed to be used with the Arduino.

My personal favorite for project enclosures is the Solarbotics S.A.F.E. series of enclosures as shown in Figure 3.52. These enclosures are available in a variety of colors, including clear, and are also available in two sizes, standard Arduino UNO and Arduino Mega. I prefer to use the Mega S.A.F.E in my Arduino projects when possible. A 60cm x 80cm copper-clad perfboard can be mounted inside the enclosure, with room to spare for a battery, etc. Any additional components such as a TFT display, power switch, rotary encoder, etc., are mounted to the sides or top of the enclosure. Many of the projects in this book were designed to fit inside a Mega S.A.F.E. enclosure.

Figure 3.52: The Solarbotics Mega S.A.F.E. enclosure

Chapter 4 • Creating Sketches and Documenting Arduino Projects

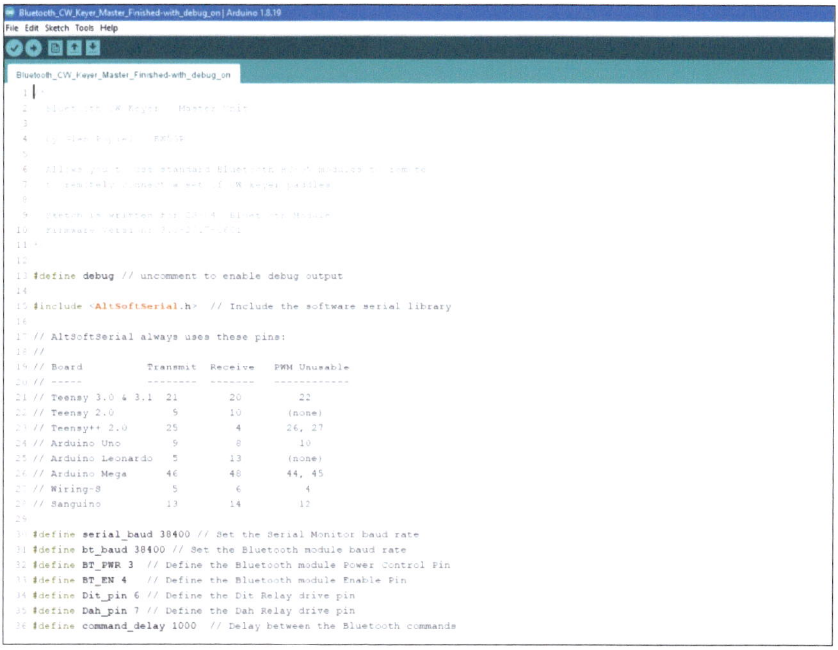

Figure 4.1: The Arduino IDE.

Arduino programs, known as sketches, are created, uploaded, and debugged using the Arduino Integrated Development Environment (IDE) shown in Figure 4.1. The IDE version used to create the sketches for the projects in this book is 1.8.19. While there are newer versions of the IDE, in particular IDE version 2.0 (2.2.1) and the Arduino Web Editor, they're just not at that level of comfort and stability that I personally would prefer at the time of authoring this book in the Fall of 2023. Feel free to use a later version of the IDE with the projects in this book and your Arduino projects in general, however, if you have any problems with a sketch from this book, please try using IDE Version 1.8.19 before any further troubleshooting. Of course, you may contact me at any time via email (kw5gp@arrl.net) for assistance.

This book is not intended to teach you how to use the Arduino IDE, as there are already a number of excellent books available to learn how to use the IDE, including *The Arduino Cookbook* by Michael Margolis (ISBN 978-1449313876), and *Programming Arduino – Getting Started with Sketches* by Simon Monk (ISBN: 978-0-07-178422-1) that can devote far more space to teaching you how to use the Arduino IDE than is possible here. There are also some excellent online tutorials for installing and learning how to use the Arduino IDE at www.arduino.cc, learn.sparkfun.com, learn.adafruit.com, as well as YouTube and other online resources.

While this book assumes that you already have a working knowledge of using the IDE, there are several important things to note as it relates to the various projects in this book, such as installing Arduino libraries, and how to add support for Arduino variant boards to the Arduino IDE.

Since this book uses a number of Arduino libraries you may not be familiar with, it is important to know how to add libraries to the IDE, as this process alone constitutes the vast majority of email I receive regarding problems with Arduino projects in my books.

Creating programs, known as sketches in the Arduino world, is part of every Arduino project. But before you even start to write the sketch for your Arduino project, it helps to plan out ahead of time exactly what you want your sketch to do, what I/O pins and methods it will use, and what shields or modules you'll be using.

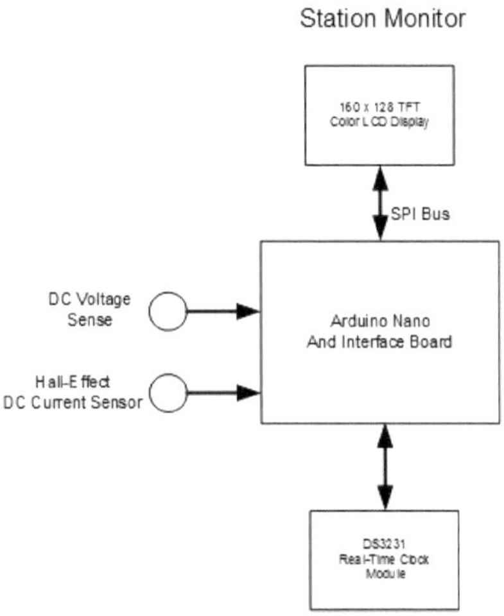

Figure 4.2: An example of a Block Diagram.

Generally, I like to start out with a "Block Diagram" of the project hardware. A block diagram similar to the one in Figure 4.2, is a very high-level overview of your project, giving you an idea of what shields, modules, and components you will need to assemble your project. A block diagram will also help identify what I/O pins and methods to use in your project. You can also use the block diagram to help determine which Arduino processor board or variant may be best suited for your project. This is also a good time to think about what type of enclosure you want to use for your project.

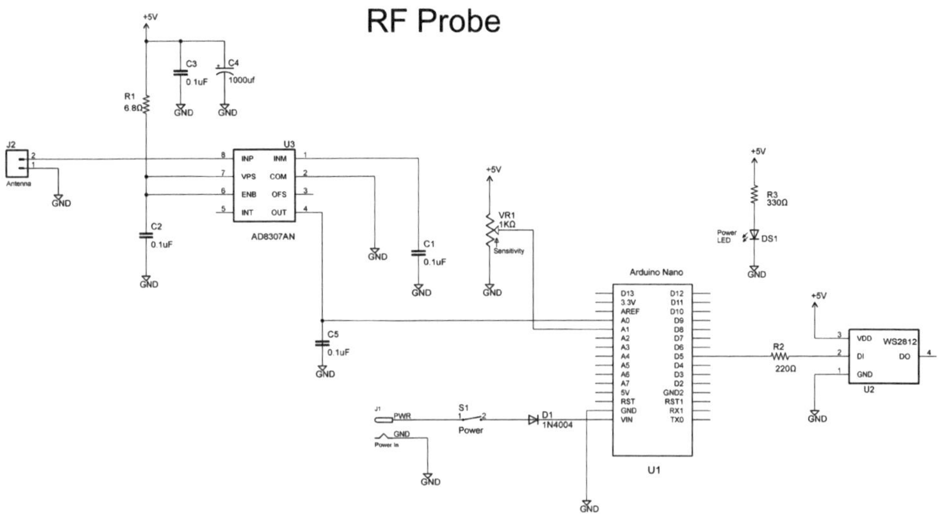

Figure 4.3: An example of a Schematic Diagram.

After I create the block diagram, I'll draw out a draft of the schematic diagram, similar to the one in Figure 4.3, using the block diagram as a rough guideline. A schematic diagram is a symbolic drawing that shows the physical connections for all of the components in your project. Assigning which I/O pins to use for the project is determined in a large part by the types of modules and components you plan to use. Since devices and modules that use the SPI or I²C bus will already be mostly assigned to their respective bus bins, all that remains is figuring out which pins to use for the remaining components. A schematic diagram can be used as a step-by-step drawing to keep track of the project build and will also help prevent wiring or design errors that you may otherwise overlook.

If you are unfamiliar with how to read a schematic diagram, there are a number of websites such as SparkFun, Circuit Basics, and YouTube that have some very good tutorials on reading schematics. Another excellent resource for learning how to read schematics is *Beginner's Guide to Reading Schematics* by Stan Gibilisco (ISBN 978-0-07-182778-2).

Finally, if I am planning on using a block of code or library for a device that was created by someone else, I'll often use the same pins they used in their project, so that I can use their sketch as a way to test that piece of the project. Once the project is complete, I'll go back and update the draft schematic diagram with the actual finished project design.

After I finish building the project prototype, I'll use small chunks of code to test the various functions of each component on the board. Once I have an assembled project that I am reasonably sure will work as intended, it's time to start work on the sketch itself. Before I actually start writing code, first I'll create something that I've used since my dear mainframe days so long ago, a Flowchart (Figure 4.4).

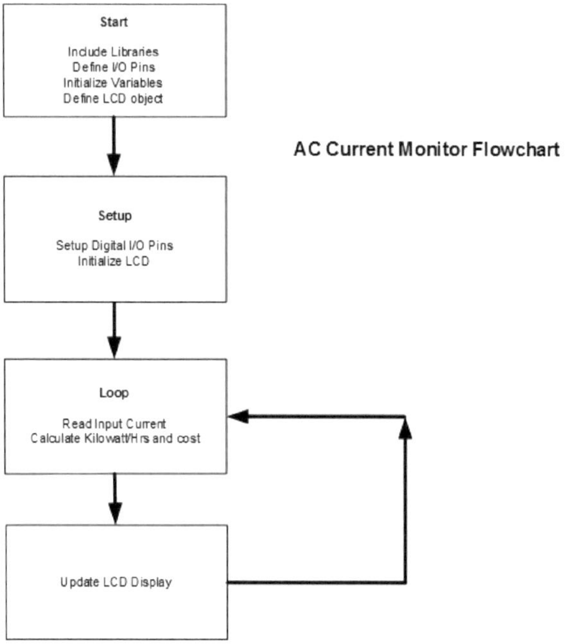

Figure 4.4: An example of a Flowchart.

Flowcharts are essentially a block diagram for your sketch. This helps to get your mind organized as to how the sketch needs to be written, what constants and variables and their type you may need, as well as helping determine what libraries you may need to include in your sketch. Flowcharting also helps to break down your sketch into smaller pieces, using the time-honored "divide and conquer method". This method allows you to write your sketch in blocks, testing each block as you go along. That way, when something goes wrong, you have a good idea of where the problem may be.

By taking the time to do this basic documentation ahead of time, you'll find that this is a great benefit when it comes time to write the sketch. It gives you a chance to review your design and see if there is anything you may have overlooked during the initial building and testing phase. Since you hopefully did some basic testing when you completed the project build, you can be reasonably certain that any problems that crop up at this point are in the sketch itself and not with the hardware. At the same time, when troubleshooting a project, especially a new one, never assume blindly that the hardware is 100% right just because it worked earlier. Wires can fall off, connectors can come loose, etc. When troubleshooting, re-verify everything that you can before thinking that the problem may lie elsewhere.

When you finish your project, it's a good idea to take the time to document it, especially if you plan to share your project with others. As you will see with each of the projects in this book, once the project is finished and tested, I use Autodesk's Eagle PCB software to create a schematic drawing and then import the finished drawing into KiCad for the final version. The reason I import the final drawing into KiCad format is part of my transitioning to a different package to do my schematic diagrams. The Autodesk Eagle package that I have used in the past is being phased out in 2026 in favor of their Fusion 360 package. While the Autodesk Fusion 360 package will still have a free version for personal use, I think it's time for me to switch over to the fully Open Source KiCad Electronics Design Automation Suite. Both packages allow you to create schematic diagrams, board layouts, parts lists, and they can even generate a file you can use to have your project etched onto a circuit board from one of the many online circuit board etching services. By transferring my finished projects into KiCad, all the components I have created, along with all of my schematics are converted to KiCad format, saving me time in the future by not having to recreate everything.

The method you choose to plan and organize your projects is a matter of personal choice, so use the method that works best for you. The important thing is to take the time to plan as much as possible ahead of time. A little bit of planning and documentation upfront can help keep you from soldering or coding yourself into a corner and getting frustrated. Remember, building things and working with the Arduino is supposed to be fun. Using block diagrams, schematics, and flowcharts can help keep things fun.

The projects in this book include a block diagram, flowchart, schematic diagram, and a parts list (bill of materials) as an aid to the construction process.

The Arduino IDE is used more than to simply create and upload sketches. The IDE has a message area that is used to provide feedback while saving, compiling, and uploading sketches, and is where any error messages are displayed. The toolbar icons near the top of the IDE window provides quick access to the most commonly used IDE functions, Verify, Upload, New, Open, and Save. The Verify operation will compile and check your sketch for errors but does not upload the sketch to the Arduino. The Upload operation does the same as a Verify, except when the sketch compiles without any errors, it will upload it to the Arduino. The New, Open, and Save operations allow you to create a new sketch, open, or save an existing sketch.

It is important to note that the Arduino IDE performs an auto-reset on the Arduino as part of the upload process. Some Arduino boards and variants do not support the auto-reset function. On these boards, you will have to manually press the Arduino's reset button immediately prior to selecting UPLOAD in the IDE. Also, Arduino boards such as the Leonardo and some variants will lose connectivity with the workstation as part of the auto-reset process, which may cause the USB port number to change when the board reconnects to the workstation. This is primarily due to a newer USB interface design that integrates the USB functionality on the Arduino (or variant) processor itself, which allows the Arduino to also act as a Human Interface Device (HID) to a connected workstation. As part of the reset process, your workstation attempts to identify the Arduino and its USB port but may fail to do so in the allowed time frame. This is a case where you may have to go into the IDE settings and reset the Serial Port selection back to the proper port. Sometimes, the Arduino's USB port doesn't show up at all after an upload or reset. The only solution I have found for this is to reset the Arduino or disconnect and reconnect the Arduino USB port until the IDE finally figures things out. This issue can get annoying sometimes and is the primary reason why I prefer the Arduino UNO or Nano over the Leonardo for many of my projects.

An interesting feature of the IDE sketch editor is the ability to create your sketch using multiple tabs from within the editor. In theory, this would allow you to break a large sketch into smaller, more readable sections, as well as giving you the ability to incorporate other files usable by the Arduino IDE such as .h, .cpp, and .c files, files that are often used within a library. In reality, using multiple tabs to create your Arduino sketch can be confusing, since the IDE will compile your sketch in alphabetical order starting with the main sketch and add, or concatenate, the tabbed files so they effectively appear to be at the bottom of the main sketch. This can cause issues with any #define statements and variables defined in the tabbed files, but used in the main sketch, among other things. Also, some developers have used this as a quick way to add a library to their Arduino sketch. While this method may work, in my personal option, it is not the best way to add a library to the Arduino IDE and your sketch, as we will see in a little bit. At the end of the day, creating a sketch using multiple tabbed files is often more trouble than it is worth, and it's best to just use a single tab for your Arduino sketch where possible.

The Tools menu is where you select the type of Arduino board you are working with, along with the COM port the board is attached to. If you do not select the correct board and COM port, your sketch may verify and compile, but the upload process will fail. You can also add support for third-party Arduino boards, by placing their board definitions, core libraries, bootloaders, and programmer definitions into a sub-folder in the hardware folder of your Arduino sketchbook. You can create the hardware folder if it does not exist, and the IDE will incorporate the new board into the IDE the next time the IDE is started.

Later versions of the IDE also have a Board Manager (Figure 4.5), similar to the Library Manager we'll talk about later in this chapter. Using the Board Manager, you can quickly and easily add support for additional Arduino processor boards to your IDE. You can also add support for additional boards by adding a third-party URL (Uniform Resource Locator) for the new boards in the IDE preferences and they will automatically be added to the IDE Board Manager, where you can select and download everything you need to support a new Arduino board or variant.

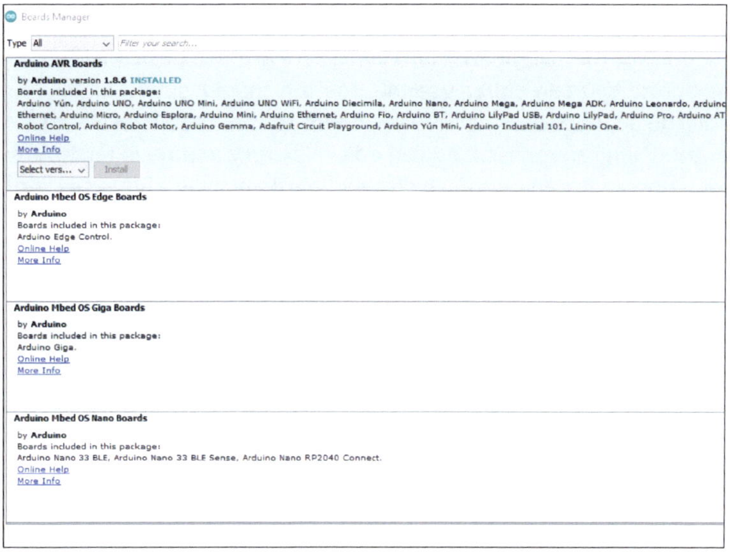

Figure 4.5: The IDE Board Manager Screen.

Also on the Tools menu, you will find the Auto Format option. This will reformat the text in your sketch so that the curly braces all line up, and the text is properly indented, making your sketches easier to read, which is a great help when you get around to troubleshooting your sketch. You can also save your Arduino sketch in compressed .zip format by using the Archive Sketch option. This will create a compressed .zip file of your Arduino sketch and save it in the sketchbook folder.

One of the most important options on the IDE Tools menu is the Serial Monitor. The Serial Monitor will display the serial data sent and received from the Arduino's USB/Serial port. Typically, you will use the Serial Monitor to show basic sketch output and display any debugging information you may have added to your sketch. You can also use the Serial Monitor to send characters and commands to the Arduino via the USB/Serial port. The Arduino's USB/Serial port can also be used to allow your Arduino to communicate with the Processing language if you have it installed on your PC, and to allow it and other PC-side applications to interact with your Arduino sketch.

One of the most underutilized tools in the IDE is the Serial Plotter. Similar in operation to the Serial Monitor, the Serial Plotter allows you to generate a real-time graphical plot of a data string that your Arduino project sends to the Serial port.

Adding New Boards

Starting with version 1.6.4, the Arduino IDE now includes a Board Manager feature that allows you add support for new Arduino and Arduino-variant boards far easier than in the past. The IDE Board Manager also keeps track of any updates and allows you to update existing board definitions as they are released.

Adding support for a new board has never been easier, and many Arduino board suppliers are adding support for the IDE Board Manager. There are two basic methods of adding support to a new board. You can either use the Board Manager option under the Tools menu, or you can add the support URL for the new board to the Additional Boards Manager URLs section under the IDE Preferences. You can edit this entry and have multiple entries, allowing you to add support for any number of new boards quickly and easily. An unofficial list of third-party URLS for the IDE is available at:
github.com/arduino/Arduino/wiki/Unofficial-list-of-3rd-party-boards-support-urls.

Your board manufacturer may also provide you with a URL for adding support for their boards that you can add to the Additional Boards Manager URLs directly.

To begin adding support for a new board, in this case the Adafruit Feather boards, you first add the support URL to the Additional Boards Manager URLs list under the IDE Preferences as shown in Figure 4.6.

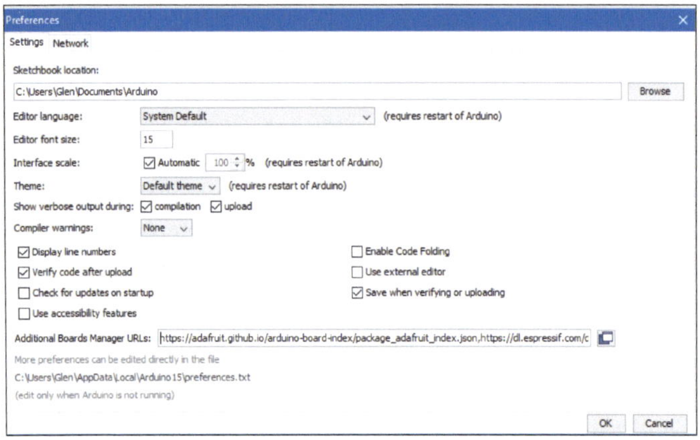

Figure 4.6: The Additional Boards Manager URL screen.

After you have added the Third-party URL to your preferences, select the Board Manager under the Tools>Board menu. When the Board Manager opens, select "Contributed" for the Type. Next, you should see the new boards listed, or listed as available to install as shown in Figure 4.7. You may have to restart the IDE for it to learn about the new boards.

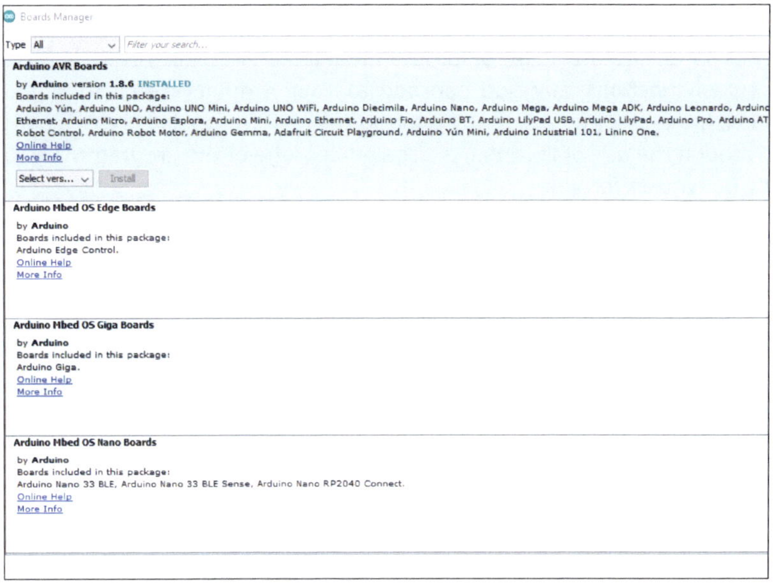

Figure 4.7: The Board Manager screen.

That's really all there is to it for adding support for new boards to your Arduino IDE. The Board Manager feature of the IDE makes adding new boards to the IDE a simple and painless process.

When you connect the new board to your Windows workstation for the first time, you may have to install a driver. In the case of the Adafruit Feather boards, you can download and run the Adafruit driver installer. Aside from clicking the default "I Agree" and "Install" boxes, the only screen to pay attention to is the screen where you are provided with checkbox options to select which board drivers to install (Figure 4.8). You can either select just the drivers for the boards you want or take the default options and install the ones recommended. As a general rule, I take the default and get them all, saving me from having to go back and repeat this process when I get another type of board from the same supplier, in this case, Adafruit.

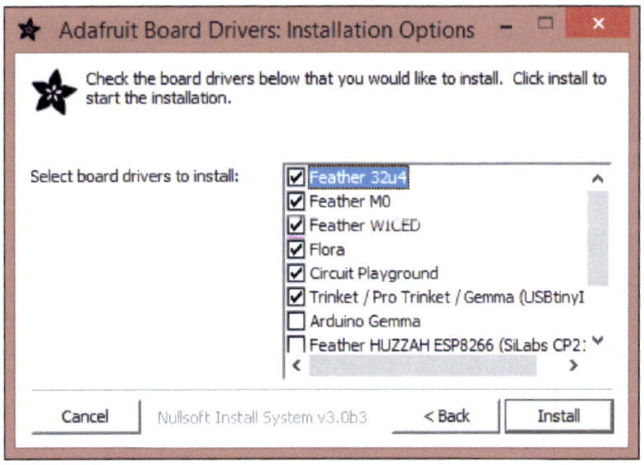

Figure 4.8: Installing the Adafruit Feather Board Drivers.

Arduino Libraries

Using libraries is an integral part of programming the Arduino. An Arduino library is a prewritten set of functions that you can add to your Arduino sketch to support shields, modules, devices, and even add additional software functionality such as trigonometry functions. Through the use of libraries, a large percentage of the programming for a project may already be written for you.

Another way to look at Arduino libraries is to think of them as the device drivers needed to interface to all of the various devices you connect to a workstation. By "including" a library in your sketch, you add the functions contained within that library without any additional programming. Then, through the use of function calls from within your sketch, you can interface the shield or device to your sketch quickly and easily. Quite often, your sketch is merely a little bit of programming "glue" you need to tie everything together, the libraries do all the hard work for you.

This is one of the major advantages to the Arduino's Open Source model. There are literally thousands of developers like you out there, creating and sharing libraries that you can use in your Arduino projects. Using Arduino libraries, you can save hours and even days of deciphering datasheets and writing test code just to interface to a new device. Many libraries even come with example code that you can integrate into your own sketches, thereby making the development and testing of your own project go much faster.

Many libraries are already integrated into the Arduino IDE, as well as a large number of other libraries available on the Internet, written and shared by their creator. In many cases, you may find that a library already exists to suit the needs of your project. More recent versions of the IDE include the Library Manager feature, which has the built-in ability to search for updates to your existing libraries, as well as allowing you to install new libraries into your IDE.

Installing Libraries

One of the biggest issues I have discovered about using the Arduino is the confusion on what a library is and how to install it. Nearly 90% of the questions I receive about problems with my Arduino book projects are library-related installation issues. If you have a sketch that uses a library, and that library is not installed, or installed incorrectly, the sketch will not compile and therefore will also not be uploaded to the Arduino. If you look at the message area on the IDE, you will see errors relating to the compiler being unable to locate functions contained within a library, along with the name of the .cpp and/or .h file associated with that library.

Libraries are not that difficult to install, and the newer versions of the IDE allow you to install libraries using several methods. The confusion regarding the installation of Arduino libraries may simply be the terminology of "installing" a library rather than what the process really involves. On the Arduino IDE, libraries are not installed in the classic sense of software installation. In a normal library installation, there is no executable file that will "install" a library for use with your IDE. Please don't overthink this, the library "installation" process is actually far simpler and more basic than it sounds.

The typical Arduino library consists of several files. In general, a basic Arduino library is written in C++ and contains a .h header file, and a .cpp file that contains the actual code for the library itself. These files are contained in the top level of a folder, or series of subfolders, with the top level folder having the same name as the .h and .cpp files it contains. A library may also contain additional files such as keywords.txt, and in the case of newer libraries that can be installed using the IDE Library Manager, a metadata file named library.properties, which contains information the IDE can use to install and update the library. A library often has additional subfolders for Example sketches and documentation on how to use the features of the library. So, essentially, at the end of the day, an Arduino library is just a group of files contained inside the Arduino sketchbook's libraries folder, in a folder with the same name as the library's .h and .cpp files.

A library can be installed manually, as a zip file through the Library Manager, or through the Library Manager itself. When you install a library, all you are doing is placing the library folder, and any subfolders that it contains, into the libraries folder in your Arduino sketchbook folder. In Windows, this is typically in your Documents>Arduino folder. This is the same folder that will contain your Arduino sketches. To manually install a library, all you have to do is copy the folder of the library you wish to use into the "libraries" folder. If the libraries folder does not exist, simply create a folder named "libraries" in your Arduino sketchbook folder and copy your libraries into this folder. In order for the IDE to see the new library, you may have to exit and restart the IDE.

At startup, the IDE will check the contents of the libraries folder, and any libraries in that folder will then be available for use in your sketches. You will see all available libraries under the Sketch>Include Library menu option. Any available Example sketches for your libraries will also be available under the File>Examples option of the IDE. All of the libraries for the projects in this book that are on the www.kw5gp.com website were designed to be installed manually, and you will need to download and extract the project compressed .zip file and then copy each individual library folder into your sketchbook's libraries folder. Figure 4.9 shows what the sketchbook libraries folder should look like when you have libraries properly installed.

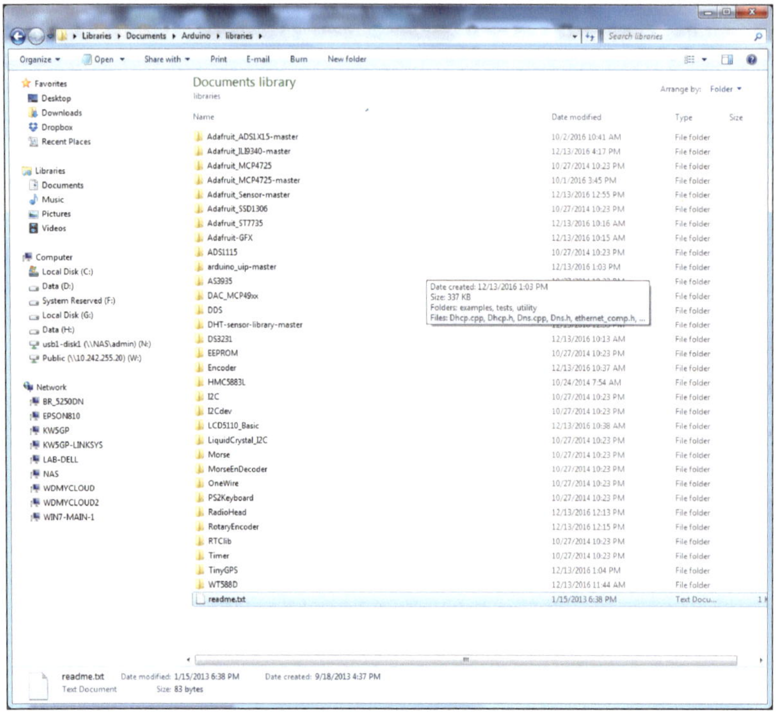

Figure 4.9: The Arduino libraries folder.

Another way to install an Arduino file is using the Library Manager's .zip file method. While many libraries have been modified to use the new Library Manager, some are still available for download as a compressed .zip file. To install these libraries, all you need to do is download them and use the ADD .ZIP Library option under the Sketch>Include Library menu. The Library Manager will extract the zipped library file and place it into the sketchbook libraries folder for you. After you exit and restart the IDE, the library will then be available for use.

Starting with version 1.5 of the IDE, a Library Manager was added to the Arduino IDE. When you open the Library Manager located under Sketch>Include Library>Manage Libraries menu option, you will see a list of libraries available for you to automatically download, install, and update using the Library Manager. All you have to do is select the library you wish to install, and the Library Manager will take it from there. As with the installation of any new library, it's always best to exit and restart the IDE after installing the library to allow the IDE to properly add the new library to the list of available libraries. The Library Manager feature is very nice and has really helped with problems related to installing libraries. Additional information on how to install Arduino libraries can be found at www.arduino.cc/en/Guide/Libraries.

Using Libraries

To use a library in your Arduino sketch, all you have to do is either select the library using the Sketch>Include Library option in the IDE, or you can manually use the library by using a #include <library.h> preprocessor definition near the beginning of your sketch. Sometimes you will see the include statement using double quotes instead of the left and right carat surrounding the library name, i.e., #include "library.h" versus #include <library.h>. For all intents, these two forms for the include statement are interchangeable, however, the double quote form allows you to have the library files in the same folder as your sketch .ino file instead of the normal libraries folder. If the IDE does not find the library in the sketch folder, it will then look in the libraries folder. The #include <library.h> form will only look in the normal libraries folder. In theory, while this does allow you to have multiple versions of a library you have customized, it's generally best to have all of your libraries in the normal libraries folder. When using a library, the include statement must always include the .h file extension as part of the included library's file name.

Some libraries include a keywords.txt file. This is a text file that will list all of the functions available in the library. This file is also used by the IDE to highlight these keywords while editing your sketch.

One final thing to remember about Arduino libraries. Many Arduino libraries are created by people just like you and me. Sometimes bugs and glitches slip in and cause things to not work as expected. I have found these library glitches to be rare, and are often quickly addressed and resolved, however, you do need to be aware that if you are having issues, be sure to check that the library functions you use in your sketch are working as expected. For example, there was an issue with the built-in Liquid Crystal I^{2}C library, where the library would only print the first letter of a string sent to the LCD. A work-around was quickly found and posted to the various Arduino forums, and the issue was fixed in a library update, but you do need to be aware that libraries can sometimes not work as intended and give you all sorts of strange problems. In many cases, there are multiple libraries available that you can use. If you suspect that you're having a library issue, try another library to be sure that is the issue.

Troubleshooting

As a general rule, the Arduino IDE is a stable and reliable platform to develop sketches for your Arduino projects. There are a few issues you need to be aware of when using the Arduino IDE, particularly with some of the newer Arduino-compatible boards and variants. With the introduction of the Arduino Leonardo and the Atmel 32u4 chip, the USB functions on the Arduino board were added to the 32u4, and the 16U2 USB controller on the original Arduino boards was no longer needed. As mentioned earlier, the USB controller in the 32u4 chip functions a little bit differently than with the UNO. When you upload a sketch to a 32u4-based Arduino, the auto-reset that the IDE performs on the board as part of the upload process can cause your workstation to briefly lose connectivity with the Arduino and when it reconnects, the Arduino may be assigned to a different USB port than what you have set in the IDE. The simple answer here is to select the new USB serial port in the IDE and continue normally. Sometimes, the board will not reconnect after uploading a sketch. This is primarily due to the USB driver on your workstation not seeing the Arduino before timing out while trying to re-establish communication. In these cases, you may have to disconnect the USB cable from your workstation to the Arduino a couple of times before everything gets sorted out.

Some of the Arduino-compatible boards use the WCH CH34x series of USB converter chips in place of the Atmel 16U2. When you connect an Arduino-compatible board that uses the CH34x USB chip, your workstation will not be able to recognize the new board until you install the CH34x USB drivers on your workstation. Locating these drivers can be somewhat troublesome, but there are more and more websites now offering the CH34x drivers for download. I was able to locate the drivers for my CH34x on the manufacturer's website at http://www.wch.cn/download/CH341SER_EXE.html. The driver for the newer CH341 will also work with the CH340 chip, so all you need is the CH341 driver file for your workstation (PC, MAC, or Linux). If you are unable to locate the drivers on the site listed above, you can find them simply by doing a web search for Arduino CH340 or CH341 drivers. Once the CH34x driver is installed, your workstation should be able to identify and connect to the Arduino-compatible board.

With some of the newer, more powerful variant boards, as well as some of the inexpensive Arduino UNO and Nano-compatible boards, I have had issues with a workstation not being able to recognize and communicate with the board, even when the correct drivers had been installed. The board worked fine when I connected it to a different workstation. As it turned out, the processor board that I was trying to use was drawing too much power from the USB port I had it connected to. I ended up having to use a powered USB Hub in order for the workstation to recognize the board.

The message area in the Arduino IDE can provide a great deal of information if you have issues compiling a sketch. In your File>Preferences menu, you can select verbose compiler output to get more detail on any compiler or upload messages, as well as setting the type of compiler warnings shown. Compiler warnings are just that, warnings to alert you to a possible issue. Many compiler warnings can be ignored, but it is beneficial to pay attention to the information provided if, and when, you see a compiler warning. As with many compilers, a compile error may not show you the exact statement that failed and why, but if you carefully read the information, it will usually provide enough information to track down the offending issue.

Memory Issues

The Arduino memory architecture is not the same as in a standard computer. The Arduino has three, and in some cases, two separate types of memory onboard. The Arduino UNO has just 32 KB of Flash memory, which is used to store the compiled sketch, 2 KB of Static RAM (SRAM) to store your program variables, data, and memory required by the actual Arduino sketch to operate properly. The UNO also has 1 KB of EEPROM that can be used to permanently store information, but few sketches actually use or need the EEPROM, and many of the variant boards do not have EEPROM.

As such, memory in the Arduino UNO and similar boards is at a premium to say the least. In general, you can fit the vast majority of your sketches in memory just fine. With the larger memory capabilities of the newer Arduino and variant boards, lack of memory is becoming less and less of an issue. It is important for you to keep the limited memory capacity of the Arduino in the back of your mind while creating your Arduino sketch. Unlike a regular workstation that will politely give you an "Insufficient Memory" error and refuse to run the program, the Arduino will do exactly as it is told and keep on running, start freezing up, doing weird things, and appearing to not execute the simplest of commands properly. There are some programs and libraries, such as the Memory Free library, you can incorporate into your sketches to keep track of available memory and let you know how much memory you have remaining.

If you find yourself needing to save precious SRAM, you can instruct the IDE to place static data and constants into Flash memory by including the avr/pgmspace.h library and using the PROGMEM keyword in your sketch. The pgmspace library allows you to store static variables, constants, strings, and arrays in flash memory instead of SRAM. You can also use the FLASH library to store string, array, table, and string arrays in flash memory. Finally, you can also use the F() syntax for storing string constants in flash. Just remember, that if you put it in flash memory, it cannot be modified during execution of the sketch, so you can only use these memory saving features for data that doesn't change.

Simple Debugging Methods

You will see in many of the sketches for the projects in this book that I have included the preprocessor directive, #define debug. You can then use a #ifdef debug and #endif to add a block of code you want to execute to your sketch to help debug the sketch. This debug code is typically used to output additional information to the Serial Monitor that can be used to monitor critical points in the sketch as an aid to troubleshooting. Once the sketch is tested and working, the #define debug statement is commented out and the debug code is ignored. This is a quick and easy way to add troubleshooting code to your sketch, and then turn it off once troubleshooting is complete.

If you use the #ifdef debug and #endif preprocessor directives around a block of trouble-shooting code, when debug is undefined (i.e., the #define debug statement is commented out), the IDE compiler will ignore the troubleshooting code and not compile it, making your finished compiled sketch smaller and more efficient.

Chapter 5 • Project Ideas, Tools, Construction, and Troubleshooting

From Dreams to Reality

Figure 5.1: The Lightning Detector Project.

Sometimes the hardest part of an Arduino project is figuring out what you want to build. The best advice I can give you here is to free your imagination and look around your ham shack. Think about what you'd like to automate or what accessories you'd like to have. Don't limit yourself to just your ham shack. You'd be amazed at some of the cool Arduino projects I've come up with that have absolutely no ham radio value whatsoever, but simply had to be built for the fun and learning experience as well as having a real purpose. Don't worry that your idea may sound too crazy or impossible, you never know until you try.

For example, using an Arduino to send the 65 tones needed for the JT65 mode on the CW-only Ten Tec Rebel Transceiver seemed like an absolutely insane idea at first thought. That crazy idea and project came from a simple question asked at my very first Arduino forum at the Huntsville, AL hamfest about 10 years ago. After a quick explanation as to what JT65 was (I had no clue what JT65 was at the time), my answer kicked off a whole series of events to make such a project a reality. A month or two later, the team of absolutely amazing minds that had been hastily put together by Craig Behrens, NM4T (SK), had cobbled together a working prototype. So, just because a project idea may sound crazy and impossible, it doesn't mean that it truly is crazy and impossible. To me, that's part of the magic and appeal of the Arduino.

So, I encourage you go to club meetings and hamfests, ask questions, attend hamfest forums, webinars, wherever you can chase knowledge and find ideas. The project idea that kicked off my whole Arduino adventure and writing career, an Arduino-powered Lightning Detector (Figure 5.1), came from a simple discussion over a breakfast meeting with my fellow members of the Olive Branch Amateur Radio Club (OBARC). It all started with these five little words, "Wouldn't it be cool if ... ?" That's actually how many of the Arduino projects I come up with begin, by asking myself that simple question, or asking my fellow hams what they think would be a cool project idea.

Browsing the "boneyard", flea market, or swap area of a hamfest is like Christmastime to me. I love digging deep through the piles of old forgotten parts and gear, or pieces of equipment that no longer function, and try to envision what I could do with them to bring them back to life or make them into something newer and better. And the brainstorming doesn't end there. I go through the vendor area and look at all the new stuff, thinking on how an Arduino could be used to improve, or duplicate, their equipment. And finally, there's no better place to catch up with old friends or co-workers as so often happens with me when I go to hamfests and pick their brain for ideas. Some of what I feel are my best ideas have come from those hamfest aisle meetings alone.

Whatever you do, keep a pen and notebook handy to write these ideas down as they come. Don't be like me and trust your memory to remember that crazy idea you came up with while rummaging through the pile of parts on a swap table. I'll see something, come up with a great idea for a project, only for that idea to vanish like a puff of resistor smoke when I see something bright and shiny on the next table.

Now that you have a list of project ideas, pick one and start fleshing it out. Is there a module or shield already out there that you can use to simplify the project build process? Don't be afraid to build on the work of others, that's the heart of what Open Source is, the sharing of ideas and information. If you're not an electronic design genius, don't worry, that's not what the Arduino is about. I'm probably the world's worst at designing analog circuits, but I do know just enough to do an Internet search to find a design that can easily be adapted to what I need. Don't be afraid, the worst that can happen is a handful of inexpensive parts goes up in smoke. It happens to everyone, no matter how good they are.

But it never hurts to learn the basics of electronic design. Naturally, the Internet is a great place to find tutorials and videos on basic electronics and electronic design. Another great place to start is the *ARRL Handbook for Radio Communications* (ISBN: 9781625951700), available from ARRL.org and Amazon.com.

I also recommend building up a supply of basic components such as resistors, capacitors, transistors, diodes, potentiometers, rotary encoders, LEDs, and the like. You'll quickly find that you'll end up using the same values of components and the same types of modules in many of your projects. Don't invest too heavily in components until you get a feel for the values and types that work best for you and the projects you design.

As for where to get the parts for your projects, that's actually simpler than one would think. With Radio Shack all but gone, the best place to find electronic components and Arduino modules is online. eBay, Amazon, AliExpress, and Banggood.com are my go-to suppliers, in addition to the mainstream Arduino suppliers such as Adafruit, SparkFun, DFRobot, and Solarbotics. When it comes to Arduino projects, parts are parts. You won't typically need the super high quality parts that NASA and the military requires, so save your money and buy the inexpensive parts when you can.

Of course you'll need some place to install all these parts. My personal choice, where possible, is to use a 60 mm x 80 mm copper-clad prototyping board similar to the one shown in Figure 5.2. I like this particular size because it fits in a Solarbotics MegaSAFE enclosure with room left over to add external components such as a 9 volt battery, TFT display, rotary encoder, power connector, switches, etc. If I'm using an Arduino UNO-footprint-type board, I like to use a prototyping shield (Figure 5.3), and a MegaSAFE enclosure. For larger projects, I usually use a larger piece of perfboard and figure out something suitable for an enclosure. If you have access to a CNC machine, 3D Printer, or laser cutter, you can go to websites such as Thingiverse and download enclosures you can make yourself.

Figure 5.2: A project built on a 60 mm x 80 mm perfboard.

Figure 5.3: An Arduino UNO prototyping shield and header pins.

I highly recommend putting your chips, Arduino boards, and modules in sockets where possible. Inevitably, you'll make a mistake and damage an Arduino board or module. Desoldering the 30 pins on an Arduino Nano is a royal pain, and after doing it a few times, you'll appreciate the value of using a socket. I create sockets for my Arduino boards and modules using DuPont-style 2.54mm spacing header sockets as shown in Figure 5.4. I also use the header pins and connector housings to create connection points to the boards and the cables connecting the various piece of my projects together as shown in Figure 5.5. For the cables themselves, I use pieces of old PC ribbon cables I pick up at hamfests and cut them to size.

Figure 5.4: Using 2.54mm SIP female socket strips to make sockets for an Arduino Nano and a Level Shifter module.

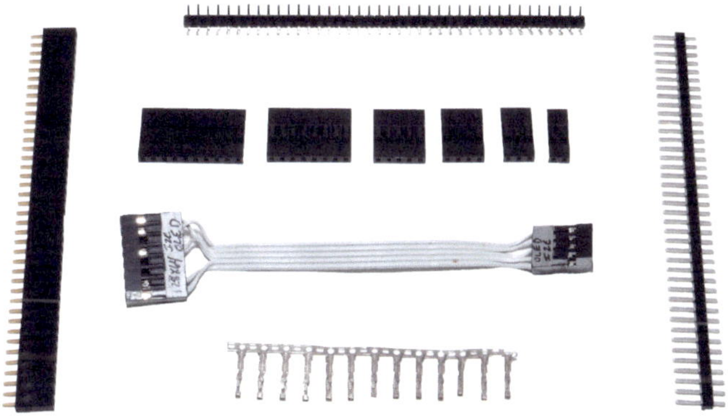

Figure 5.5: The various 2.54-mm (0.1 inch) options and an interface cable built from header pins and connector housings.

To wire everything together, I like to use the bottom of the board to keep the ugly mess of wires out of sight. Using 30 gauge Kynar™ solid-core wire-wrap wire, I connect all the pieces of the project together. The wire is available in many colors, allowing you to color-code the power, ground, and signal wires as shown in Figure 5.6.

Figure 5.6: Wiring of the Waveform Generator project.

Finishing Touches

Inevitably, you'll want to place your finished project into an enclosure of some sort. As I have mentioned before, I really like using the Solarbotics SAFE series of enclosures when possible. Not only does an enclosure protect your project from damage, it adds a little extra something to the looks. Interestingly enough, craft stores such as Hobby Lobby also have some interesting options for enclosures, and I often use their plastic golf ball, baseball, and softball trophy cases as Arduino project enclosures.

Of course, once you have your project in an enclosure, you'll probably want to label the various connection points and controls on the enclosure. I've found a really nice, professional-looking way to add labels to my finished projects.

You can use water-soluble inkjet or laser-printable decal sheets to create colored labels, logos, and other types of artwork for your project enclosures. These decal sheets are similar to the decals used on the plastic models many of us built as kids. All you have to do is print the image on the decal paper, making sure that you use the right type of paper for your printer. After printing, the laser-printed decals are ready to go, the inkjet-printed ones need the extra step of spraying a sealer on the printed label to keep the ink from running when it gets wet. Figure 5.7 shows an example of a 40-meter Arduino-powered CW transceiver using this type of decal as a logo on the side of the Hobby Lobby baseball trophy case that I used as an enclosure.

Figure 5.7: 40-meter CW transceiver project showing labels and decals.

If you use one of the schematic editing packages such as KiCad, another nice finishing touch you can do is to have a professionally etched circuit board by one of the online circuit board services and custom-build an enclosure to really spice up a finished project.

Tools

So what tools will you need to build all those cool Arduino projects you have listed in your project idea notebook? As with any hobby, you can either spend a little, or spend a lot on your tools and test equipment. When it comes to the Arduino, you really don't need the best of everything to build Arduino projects, in fact, a lot of the time, you don't even need every tool in the box. As your project building skills progress, you'll naturally upgrade your tools and test equipment to suit your specific needs and tastes.

First, you'll need a good temperature-controlled soldering iron or soldering station similar to the one shown in Figure 5.8. Be sure to get one with interchangeable tips. I generally use the smallest tip available when I work with the very close connections on the bottom of the prototyping boards. You'll also want a soldering station that has a hot air gun. You'll really like having this if you ever want or need to solder or desolder surface mount components. It also does a great job with heat-shrink tubing.

Figure 5.8: A typical soldering station with heat gun.

In conjunction with a soldering station, I have found an inexpensive component tester (Figure 5.9) to be an absolute necessity. It seems that with each passing year, the color codes on the resistors get thinner and thinner, the colors are almost impossible to determine, and the writing on the components gets smaller and smaller. No matter what the truth is behind this perception, and even if my eyesight were as good as it once was, I'd still have issues determining color codes and the numbering system on capacitors.

Thankfully, the component tester solves all of these issues and more. This simple, inexpensive device will not only test a wide array of components from resistors, capacitors, coils, diodes, transistors, and more, but it will also provide a pinout of the component under test, along with some other basic data about the component. It's also great at determining just what an unmarked component actually is. Newer versions of the component testers will even display a schematic symbol of the component with the pin numbers appropriately displayed. This simple device is probably the most used tool in my work area. It's hard to imagine building projects without it.

Figure 5.9: A component tester.

Naturally, I highly recommend getting a standard multimeter similar to the one shown in Figure 5.10. The one pictured here is one I've had for many, many years, but in actuality, the inexpensive ones from places such as Harbor Freight work every bit as well, and even have more functions that my trusty old multimeter.

Figure 5.10: A typical multimeter.

While you don't need an oscilloscope to build and troubleshoot Arduino projects, a scope can definitely make life easier at times, and eventually you'll probably want one. Modern digital storage oscilloscopes (DSOs) are not that expensive anymore. Amazon and other online stores such as AliExpress have scopes similar to the one shown in Figure 5.11 for under $180 US. Of course, you can always find an older-style oscilloscope in many hamfest flea markets at a very good price.

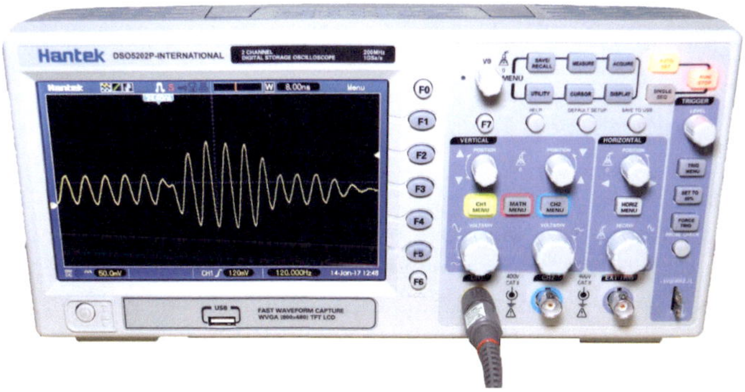

Figure 5.11: A typical digital storage oscilloscope.

Another tool you might find handy is a logic analyzer. This instrument will allow you to see the signals and timing relationships between various digital signals. You can think of it as a form of oscilloscope for digital signals only. The instrument shown in Figure 5.12 will display up to 16 separate digital signals on a PC connected via the analyzer's USB cable in a format similar to the output shown in Figure 5.13.

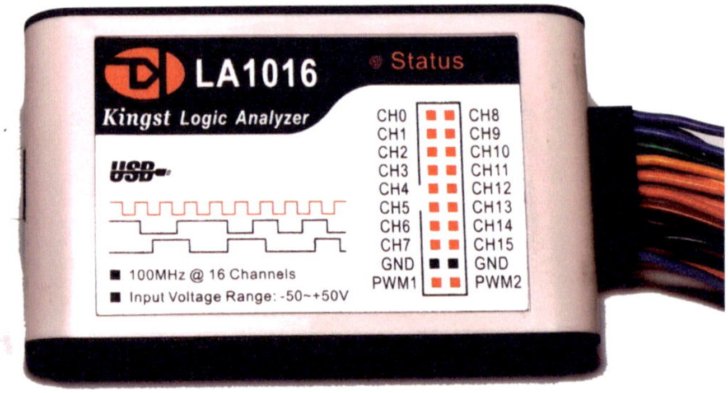

Figure 5.12: A typical logic analyzer.

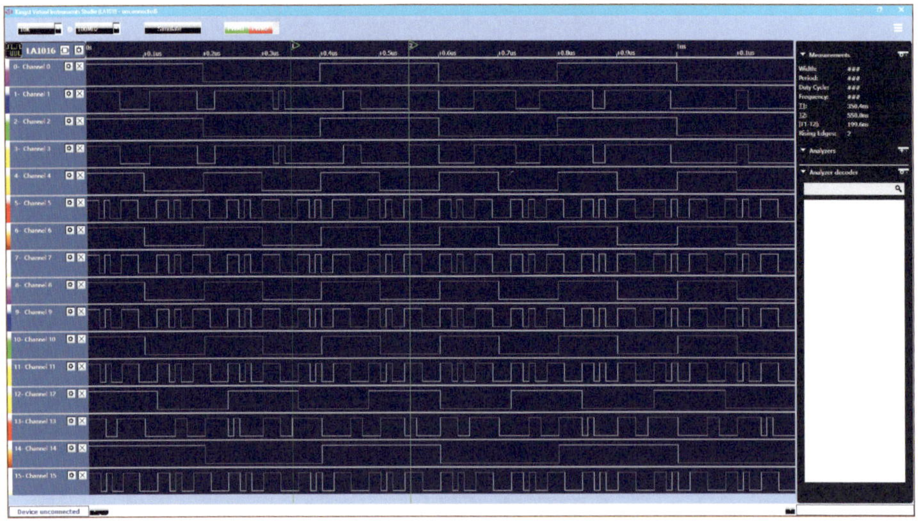

Figure 5.13: The output display of a logic analyzer.

Breadboard and Development Systems

As you start out with the Arduino, you'll most likely use a basic breadboard similar to the homebrew one I built when I first started with the Arduino (Figure 5.14). This platform served me well for many years of Arduino project development. It provided quick and easy access to the different Arduino board types, along with several of the standard modules that I would typically use with my Arduino projects. This is a great way for you to start out with prototyping your Arduino projects as well.

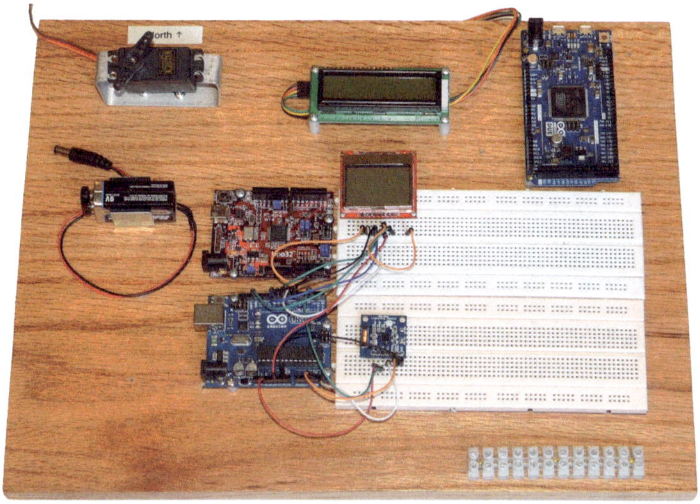

Figure 5.14: an example of a breadboard Arduino development system.

One powerful tool in my project development arsenal is the Arduino I/O shield shown in Figure 5.15. This shield brings every I/O pin on either an attached UNO or Nano out to a header pin, along with +5 volts and Ground. This allows you to use jumper wires between the I/O shield and the various modules on your breadboard or development system to create a prototype or "proof of concept" circuit. It also provides a great place to measure signals on the I/O pins with your multimeter, oscilloscope, and/or logic analyzer.

Figure 5.15: An Arduino UNO/Nano I/O shield.

Naturally, you can expect the development tools from my early Arduino days to progress as well. Recently, I switched my prototyping, test, and proof of concept development system over to several new products that I highly recommend. The first of these is the Universal WorkBench system from Phase Dock (Figure 5.16). I often refer to the WorkBench as a "breadboard system on steroids". Using a unique system of platforms known as "clicks" and "slides", the WorkBench allows you to customize your development environment quickly and easily, simply by choosing the "click/slide" assemblies you need for the project you have in mind and "clicking" them into place on a pegboard-style base. You can even get a plastic cover option to protect your project between work sessions.

Figure 5.16: The Phase Dock WorkBench development system.

I often get asked "What is the best way to start out with the Arduino?" When I began working with the Arduino, my answer was usually to go to eBay or a similar online supplier and buy an "Arduino starter kit". Unfortunately, many of the Arduino "starter" kits contain a lot of components that don't really have much application in ham radio projects. The second problem with this kind of starter kit is that unless you have a breadboard setup similar to the one I started with, everything tends to get spread out on a table, leaving a mess to tear down and reconnect for every Arduino session.

To address this issue, Dr.Duino has created some great Arduino Development systems. As a starter kit, the Dr.Duino Pioneer Edition (Figure 5.17) is pretty hard to beat. The Pioneer is actually a shield for the Arduino UNO that has many basic Arduino project components mounted on the board to help you learn the Arduino and create basic Arduino projects. The Pioneer has 3 LED outputs, 3 potentiometers for analog inputs, 4 pushbutton switches, a piezo buzzer, and an 8 LED addressable RGB LED strip, along with connections for other sensors and devices such as a servo, an ultrasonic sensor, a light-dependent resistor, and a passive IR sensor, all on a shield that fits the UNO-type boards. The Pioneer comes as a kit, so you even get some soldering practice building a relatively simple but very versatile Arduino starter system. Since everything is on one board, there's a lot less stuff put away when you're done for the day. Dr.Duino has set up an exclusive special Pioneer offer for readers of this book at http://www.drduino.com/Glen-Pioneer.

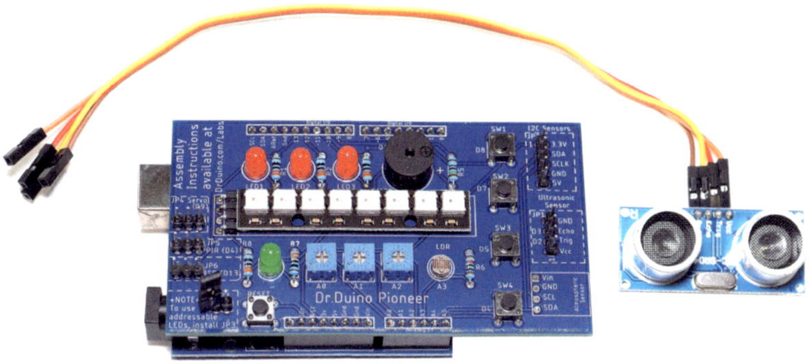

Figure 5.17: The Dr.Duino Pioneer Edition.

As your skills progress, you might consider moving up to the Dr.Duino Explorer Edition (Figure 5.18). The DrDuino Explorer is targeted to be a mini-development platform. The Explorer has two breadboard areas; the standard temporary wire jumper type, and a second area with solder pads for more permanent uses.

The Explorer features four LED outputs, 4 pushbutton inputs, 3 potentiometers for analog input, a light-dependent resistor, an HC-05 Bluetooth module, a 128x32 Organic LED display, a 5 V / 3 A DC-to-DC converter, and an 8 RGB addressable LED strip, along with additional headers for external connections for an atmospheric sensor, an ultrasonic proximity sensor, a servo, and other external device connectors. The Explorer also has all of the Arduino I/O pins brought out to jumpers on the board, allowing you to re-route one or more I/O pins to the Explorer on-board components or pass these signals on to an UNO-type shield attached to the Explorer board. The Explorer has header sockets to allow it to be used with either an Arduino UNO or Nano processor board. For me, the Explorer has become my go-to platform for building and testing my "proof of concept" ideas and for prototyping Arduino UNO and Nano projects. Dr.Duino has set up an exclusive special Explorer offer for readers of this book at: https://www.drduino.com/Glen-explorer.

Figure 5.18: The Dr.Duino Explorer Edition.

At some point, you'll probably start working with some of the higher-powered Arduino variants such as the STMicroelectronics STM32, Raspberry Pi Pico, and Espressif ESP32. The Dr.Duino Inventor Edition (Figure 5.19) is an Arduino-compatible development system based on the Espressif ESP32 microcontroller.

The Inventor Edition features an MP3 player module and speaker to play audio files from an SD memory card, a DHT-11 Relative Humidity and Temperature Sensor, a 128x64 Organic LED (OLED) Display, a Light Dependent Resistor (LDR), a stepper motor, three potentiometers, three pushbutton switches, 16 addressable RGB LEDs arranged in a 4 by 4 matrix, a relay, and a breadboard area. A level shifter is incorporated on the board to support the Inventor's hardware, as well as providing a few extra level shifting pins for your own use.

The Inventor also has sockets on the board that allow the mounting of a standard Arduino shield. This allows you to use the Inventor to initially create a project prototype, then move it to a more permanent solution such as a prototyping shield when you're ready. There's also a connector on the board that allows you to attach an external addressable RGB LED strip.

As with the Explorer for the UNO and Nano, the Dr.Duino Inventor has become the first thing I reach for when I'm working on a project that requires more horsepower and I/O than the UNO and Nano can provide. Dr.Duino has set up an exclusive special Inventor offer for readers of this book at: http://www.drduino.com/Glen-Inventor.

5.19: The Dr.Duino Inventor Edition.

Working with the Arduino and External Power

As you create Arduino projects, at some point you will need to troubleshoot your project while powering the Arduino from an external power source connected to the External Voltage (Vin) pin on the Arduino. If you connect the USB cable from your workstation to the Arduino while powering the Arduino externally, you can create a power conflict situation between the power on the USB connector and the external power to the Arduino. To prevent this, you can customize a USB cable with the voltage wire cut, allowing only the USB Data + and Data – signals along with the ground connection to attach to the Arduino's USB port. This simple solution will prevent any voltage conflict issues and still allow you to upload sketches and use the Serial Monitor to troubleshoot the project.

Chapter 6 • Arduino I/O Methods

Figure 6.1: Attaching a TFT display to the Arduino SPI bus.

The Arduino has a variety of ways that it can communicate with attached modules, devices and the outside world. It is important to understand how each of these methods can be used in your Arduino projects. Knowing these methods will help provide you with the information you will need in order to understand what is going on inside your Arduino project, as well as understanding how the various projects in this book function. This will also help you to add your own features and enhancements, and aid in the design and creation of your own Arduino projects.

The fundamental purpose of a microcontroller such as the Arduino is to interface directly with sensors and devices that allow the Arduino to sense and control things. The primary methods of I/O on the Arduino are digital I/O, digital I/O with Pulse Width Modulation (PWM), analog input, and TTL Serial/USB serial communication. The Arduino also supports several bus-type protocols, including the Serial Peripheral Interface (SPI), Inter-Integrated Circuit (I^2C), and 1-Wire buses. In addition to these methods, many of the newer Arduino variants support Bluetooth and WiFi wireless communication. Through the use of a shield or external module, your Arduino can also interface and communicate with USB devices. The Arduino also supports hardware and software "interrupts", which allow program execution flow to be changed (interrupted) by an external event.

Digital I/O

The simplest and most common form of I/O on the Arduino is done via the digital I/O pins. Digital I/O pins can be used to turn on LEDs, relays, and other external devices, as well as sensing the position of a switch, pushbutton, or other form of on-off (digital) input. The

Arduino digital I/O pins also have an internal pull-up resistor (normally disabled) that can be enabled via software, saving you from having to add an external pull-up resistor when sensing the position of a switch, for example. Some Arduino variants also have this internal pull-up resistor, while some do not. Be sure to verify whether or not the Arduino variant you are using supports this feature when designing your project. On the Arduino UNO, Mega2560, and Leonardo, the value of the internal pull-up resistor is from 20 kΩ to 50kΩ. The Arduino digital I/O pins are defined as being tri-state, meaning that the I/O pin can either be a logic-level High, Low, or in a high-impedance (disabled) state, which effectively isolates the I/O pin from the circuit.

Through the use of several Arduino libraries, you can also use digital I/O pins to allow TTL serial, SPI, or I^2C communication using pins other than the pins normally designated for these communication protocols.

Digital I/O with Pulse Width Modulation

Six of the digital I/O pins (pins 3, 5, 6, 9, 10, and 11) on the Arduino UNO can be configured to output data using Pulse Width Modulation (PWM). PWM allows you to control the duty cycle of a square wave output on a digital I/O pin using an 8-bit value from 0 to 255. Using PWM, you can dim an LED, output an audio tone, control the speed of a DC motor, and perform other tasks where you need to modify the duty cycle of the output waveform. The Arduino Leonardo has seven PWM-capable digital I/O pins and the Arduino Mega2560 has 14 PWM-capable digital I/O pins. The number of PWM-capable digital I/O pins varies between Arduino boards, so be sure to check the digital I/O pin specifications for the Arduino board that you plan to use in your project.

Analog Input

The Arduino UNO has six analog input pins. Even though it uses the same Atmel ATmega328 processor as the UNO, the Nano interestingly has 8 analog input pins. These pins convert an analog input of 0 to 5 Volts into a 10-bit digital value from 0 to 1023. Keep in mind that on a 3.3 Volt Arduino, the maximum voltage on the analog input pins is 3.3 Volts instead of 5 Volts. The range of the A/D converter can be modified either by using an external reference voltage or by selecting the UNO's internal reference of 1.1 Volts. The Arduino Mega2560 also supports an internal reference voltage of 2.56 Volts. The Arduino Leonardo and Mega2560 each have 12 analog input pins.

The Arduino analog input pins can also be used as digital I/O pins, and like the regular digital I/O pins, also have an internal pull-up resistor (normally disabled) that can be enabled via software. The value of the internal pull-up resistor on the analog pins is the same as for the standard digital I/O pins.

Analog Output

While the "standard" Arduinos such as the UNO, Leonardo, and Mega2560 do not have I/O pins to convert from a digital representation to an analog output voltage, some of the more recent Arduino processor boards have digital-to-analog (D/A) pins. Using the D/A pins, you can output an analog voltage, sine wave, and other waveforms directly from the Arduino itself, rather than having to use an external D/A module.

Serial I/O

Figure 6.2: The Solarbotics Ardweeny. Note that the row of 6 pins on the top right of the board are used to connect to an FTDI module for programming.

The Arduino Integrated Development Environment (IDE) on your workstation communicates with the Arduino via the hardware serial port. Most Arduinos communicate with the IDE via the USB connector, however, some Arduinos, such as the Solarbotics Ardweeny (Figure 6.2), require a USB-to TTL serial converter or "FTDI module" (Figure 6.3) to access the serial port on the Arduino. The USB port normally shares the I/O pins with the Arduino's hardware serial port. On the Arduino UNO and similar Arduinos, pins 0 and 1 are used for serial communication. When designing your Arduino project, care must be taken if you use these designated serial I/O pins in your project, as it can interfere with your ability to communicate and upload sketches to the Arduino.

Figure 6.3: An "FTDI" module.

The IDE can be used to upload sketches and the IDE also has a Serial Monitor feature that allows you to view the serial port data, as well as sending serial data to, and from, the Arduino. If your project requires a separate serial I/O port, you can use the Software Serial library, which allows you to assign any pair of digital I/O pins as a software-driven serial port. Additionally, some Arduinos and variants, such as the Mega2560 have additional hardware serial ports onboard.

1-Wire Bus

The 1-Wire Bus Interface was designed by Dallas Semiconductor Corp. (now part of Maxim Integrated Products) to provide low-speed data signaling and power over a single data line. Typically used to communicate with small devices such as temperature, voltage, and current sensors, along with external memory and other devices, the 1-Wire bus interface is a bus-type interface architecture that is implemented using a single wire (two if you count the ground wire). With a data rate of up to 16.3 Kb/s, the 1-Wire bus can communicate reliably with devices over 100 meters away from the host.

Each 1-Wire device has its own unique 64-bit serial number embedded in the device, allowing many 1-Wire devices to be attached to the same digital I/O pin, without the need for any additional configuration or wiring. Using an advanced algorithm, the host, also known as the bus master, can quickly identify all of the devices attached to the 1-Wire bus, and also determine their type based on the device type information that is embedded in the lower 8 bits of each device's serial number. This algorithm can scan the one the 1-Wire bus and identify up to 75 sensors per second.

A unique feature of the 1-Wire interface is that many 1-Wire devices can be powered entirely from a single digital I/O pin used as the data line for the 1-Wire bus. Using this method, known as "parasitic power", the device does not need a separate power supply, and instead, a small capacitor (typically 800 pF) is integrated inside the device to store enough of a charge to operate the device's interface to the bus.

MaxDetect also makes a series of relative humidity and temperature sensors that use a proprietary 1-Wire interface. The MaxDetect 1-Wire interface, however, is not a bus architecture, and the MaxDetect devices do not have embedded addresses, meaning that you can only have one MaxDetect 1-Wire device attached to the digital I/O pin at time. The MaxDetect 1-Wire interface does not support parasitic power mode and must be supplied power on a pin separate from the device's data pin. The MaxDetect 1-Wire interface is not compatible with the Dallas Semiconductor/Maxim 1-Wire bus, so care must be taken when mixing these two types of 1-Wire devices in your projects.

Both types of 1-Wire devices and interfaces are supported by Arduino sketch libraries and example sketches, which makes interfacing and using 1-Wire devices in your Arduino projects simple and easy.

Serial Peripheral Interface (SPI) Bus

The Serial Peripheral Interface (SPI) bus protocol was developed by Motorola to be a high-speed, full-duplex communication, bus-type protocol between one master and multiple slave devices. SPI, along with I²C, are essentially the "workhorse" buses that you will use most often with your Arduino projects.

The Arduino communicates with the SPI devices on a single shared bus using four signal lines. These lines are designated Clock (SCLK, SCK, or CLK), Slave or Chip Select (SS or CS), Master-Out Slave In (MOSI), and Master-In Slave Out (MOSI or DI). Each device attached to the SPI bus requires a separate Slave/Chip Select line.

SPI is a loosely defined standard and can be implemented in slightly differing ways between device manufacturers. Since SPI is considered to be a synchronous communications protocol, data is transferred using the Clock line, with no formally defined upper limit on speed. Some SPI implementations can run at over 100 Mb/s. SPI has four defined modes (Modes 0, 1, 2, and 3) which define the clock edge on which the MOSI line clocks the data out, the clock edge on which the SPI master device samples the MISO line, and the clock signal polarity. Fortunately, the Arduino sketch libraries for SPI and the various SPI devices handle the proper signaling required to communicate with the various SPI devices you may attach to your Arduino project.

On the Arduino UNO, the SCLK, MISO, and MOSI pins are defined as pins 13, 12, and 11, respectively. Digital I/O pin 10 is often used at the Slave Select pin for the first SPI device, however any available digital I/O pin may be used as a Slave Select pin. On the Arduino Mega2560, the SCLK, MOSI, and MISO pins are defined as digital I/O pins 52, 51, and 50 respectively, with SS for the first SPI device typically assigned to digital I/O pin 53. On some Arduinos such as the UNO R3, Leonardo, and Mega2560, among others, the SPI signals are also brought out to a six pin header also called the In-Circuit Serial Programming (ICSP) header.

The Arduino Leonardo does not have any digital I/O pins specifically assigned for SPI communication. Instead, the SPI signals are brought out to the ICSP header only. Since each SPI device attached to your Arduino requires a digital I/O pin assigned to the Slave Select of each SPI device, your projects are limited to the number of digital I/O pins available. Some Arduinos and variants have multiple SPI buses onboard.

There are also several Arduino libraries available that allow you to software-define the SPI bus interface to use regular digital I/O pins and communicate with the SPI devices via these pins by using software-based timing of the SPI signals (also known as bit-banging). With this method, the digital I/O pins are turned on and off manually by the library to simulate the hardware timing needed to communicate with the SPI device. While this "bit-banging" method does work, it requires all of the bit timing and clocking to be performed in software, which is much more processor-intensive and less efficient than the hardware-based SPI method. It's usually better to design your projects to use hardware SPI and only use software-based SPI communications when absolutely necessary.

As compared to the I²C bus, SPI is generally faster, and assuming you have the digital I/O pins available in your project, is the preferred method for interfacing most modules and devices that support both SPI and I²C communications.

Inter-Integrated Circuit (I²C) Bus

The Inter-Integrated Circuit (I²C) bus was developed by Phillips (now NXP Semiconductors) for attaching low-speed peripherals to a host device. On the Arduino, the I²C bus is also known as the Two-Wire Interface (TWI) bus. I²C is a serial bidirectional 8-bit communications protocol used by many manufacturers who develop peripherals and devices for embedded systems and microcontrollers such as the Arduino.

The I²C bus requires only two communication lines, Serial Data (SDA) and Serial Clock (SCL), plus power and ground. On the Arduino UNO, these are defined as analog pins A4 and A5, respectively. On the Arduino Leonardo and Mega2560, SDA and SCL are assigned to pins 20 and 21, respectively. Some Arduino variants have a second I²C interface, with the pins designated as SDA1 and SCL1.

The I²C standard defines the speed of the I²C bus as 100 Kb/s (Standard Mode), 10 Kb/s (Slow Mode), 400 Kb/s (Fast Mode), 1 Mb/s (Fast Mode plus), and 3.4 Mb/s (High Speed mode). The Arduino I²C bus defaults to a bus speed of 100 Kb/s, but the bus speed can be changed by modifying the internal Arduino Two Wire Bit Rate Register (TWBR), or by modifying the TWI speed definition in the Arduino Wire library. Unless your project requires it, and your devices can support it, it is best not to modify the Wire library, as it may cause issues when you compile other projects that use the same Wire library.

I²C devices have a unique 7 or 10-bit address, with some devices capable of having their I²C address reassigned using jumpers, switches, or some other hardware method, allowing you to have multiple devices of the same type co-existing on the I²C bus. On the Arduino, I²C addresses 0-7 and 120-127 are reserved, leaving 112 7-bit addresses available for devices. Every I²C device connects to the bus using open-drain (the same as open-collector except for MOSFET devices), requiring the use of pull-up resistors on the SDA and SCL bus lines. Typically, the value of these pull-up resistors is 4.7 kΩ, so the Arduino's digital pin's I/O internal pull-up resistor will not suffice as a viable I²C bus pull-up resistor value.

As with software SPI, there are several Arduino libraries available that allow you to use regular digital I/O pins for the I²C bus and communicate with I²C devices on those pins using software-based timing of the I²C signals (also known as bit-banging). With this method, the digital I/O pins are turned on and off manually by the library to simulate the hardware timing needed to communicate with the device. While this method does work, it requires that all of the bit timing and clocking be done in software, which is far more processor-intensive and less efficient than the hardware-based I²C method. It's far better to design your projects using hardware I²C and use software I²C communication only when absolutely necessary. Rather than using software to create a second I²C bus if needed, I would recommend using an Arduino that has additional I²C bus interfaces available onboard.

Bluetooth Communication

Figure 6.4: The HC-05 Bluetooth module.

Using Bluetooth technology with the Arduino is an often misunderstood and frustrating endeavor. Deciphering how Bluetooth technology can be integrated into your Arduino projects could easily take an entire book unto itself. It took me the longest time to get comfortable enough with the Arduino and the Bluetooth modules to where I felt I could include a Bluetooth project in a book.

Usually, when you think of Bluetooth, one typically envisions the Bluetooth devices we use in conjunction with our smartphones, such as wireless headsets and microphones, keyboards, mice, etc. These types of Bluetooth devices are often referred to as Human Interface Devices (HID).

In reality, the vast majority of Arduino projects involving Bluetooth communication use a non-HID version of Bluetooth, known as Serial Port Profile (SPP). Some of the more recent Bluetooth modules can emulate the role of an HID device when communicating with a PC or smartphone, but you cannot connect a Bluetooth keyboard to your Arduino using one of these Bluetooth modules. So, when you think of Bluetooth and the Arduino, think along the lines of a wireless serial data connection from one Bluetooth module to another, or a Bluetooth Module to a PC or smartphone. If you want to use a Bluetooth keyboard with an Arduino, more than likely the easiest way would be to use a USB Host shield/module with a standard Bluetooth dongle and connect the Bluetooth keyboard that way.

To complicate matters even further, there are multiple versions of Bluetooth technology; Bluetooth and Bluetooth Low Energy (BLE), also known as Bluetooth 4.0. Both operate in the 2.4 GHz Industrial, Scientific, and Medical (ISM) band, but that's pretty much where the similarities end.

The older Bluetooth technology is often referred to as Classic Bluetooth, and incorporates the Bluetooth 1.x, 2.x, and 3.x standards. Often, you will hear the term "EDR" (Enhanced Data Rate) associated with this version of Bluetooth. EDR refers to enhancements within this version of the Bluetooth protocol allowing faster data rates, up to 3 Mbit/s. At 1 Mbit/s, Bluetooth BLE has a slower data throughput than Classic Bluetooth, but operates at a much lower power level. Unfortunately, the two technologies don't often play well together. While a Bluetooth BLE smartphone can link and communicate with a Classic Bluetooth device, the reverse is not true. So for simplicity's sake, when working with Bluetooth devices, I recommend using the same version on both ends of the connection to avoid any compatibility issues.

When using the Bluetooth Serial Port Profile, one end of the connection is designated the Master and the other end is designated the Slave. The HC-05 Bluetooth module (Figure 6.4) is the most commonly used Bluetooth module with the Arduino. While the firmware varies widely between the HC series of modules, typically the HC-05 can be configured for either the master or slave role. The companion HC-06 module can only operate in the slave role. The HC-05 does not support the HID Bluetooth profile and therefore can only be used for serial data links. And the fun's not over yet, there is a lot of variance in the firmware for the various HC-05 modules used with the Arduino, so you really need to play with these modules to get a feel for how to integrate them into your projects.

Many of the HC-05-type modules can be powered by 5 Volts, but the signal lines operate at 3.3 Volts. To prevent damage to the module, you will need to use a voltage divider or level shifter on the signal connections to the module.

The HC-05 has two operating modes, Data and AT Command mode. The AT command mode is reminiscent of the old dial-up modem "AT" command prefixes. The command mode is used to perform any configuration required to communicate with the slave unit. The HC-05 powers up into the default Data mode and there are differing methods of entering the AT command mode based on which version of firmware you are running. The most recent and most common version of the HC-05 board has a small pushbutton switch that will put the HC-05 into AT Command mode if the switch is pressed before, and while, the unit is being powered up. The LED on the HC-05 board should blink slowly on and off, about once every two seconds to indicate that it is in AT Command mode. When in Data mode, the LED blinks faster, about 5 times a second while waiting for a connection, and then a quick double flash about every two seconds when it is linked to another Bluetooth device. You can simulate the pressing of the pushbutton switch by using a data pin driving a transistor to enable the power while a second I/O pin is used to output a logic-level "High" to the Enable (EN) pin of the module. An example of this technique can be found in the Bluetooth CW Keyer project in this book.

There are other Bluetooth modules becoming popular for use with the Arduino, including the HC-10, HC-11, RM-42, and others. Again, it would take a whole book just to cover all of the various Arduino Bluetooth modules and their individual intricacies. But once you get the hang of working with the HC-05, working with the other Bluetooth modules should be just as easy, if not easier.

USB Communications

In many ways, USB technology is like an iceberg. At the surface, it appears to be just a simple straightforward high-speed communication method. But it's that hidden 90% that will absolutely ruin your day, especially when it comes to the Arduino. There are two basic components of a USB connection, the USB Host and the Device. This distinction is an important piece in understanding USB communication as it relates to the Arduino. In order to be programmed using the Arduino IDE, just about every Arduino communicates with your PC via the Arduino's onboard USB port (a handful of Arduino-variants use only Bluetooth or WiFi), or through an FTDI module to convert between the Arduino's serial port to the PC's USB port. In these types of connections, the PC is the USB Host and the Arduino is the Device. Most USB devices are Human Interface Device (HID) such as keyboards, mice, external hard drives, etc. Here's where that dangerous unseen portion of the iceberg comes into play.

The reason for this issue is the nature of the USB protocol. USB stands for Universal Serial Bus, meaning that it is designed to be the standard for interfacing all manner of devices to the host device. As such, determining (also known as enumerating) what device has been connected, as well as installing and/or configuring the device drivers needed to communicate with the device, is necessary for the attached device to operate properly. Because of the huge variety of USB devices, you can see that setting up to communicate with a USB device is not really all that simple, but this all happens behind the scenes on your PC or smartphone. It can take a lot of processing overhead, horsepower, and memory to handle everything that is needed to communicate with a USB HID device.

Some Arduino boards can emulate a USB HID device to a host such as a PC or smartphone. But, if you want to connect a USB keyboard, mouse, or other HID device to your Arduino project, the Arduino has to have USB Host functionality. There are a few Arduino boards, such as the Arduino Due and Mega ADK, both of which are officially discontinued but are still available through suppliers such as eBay that can function as a USB host without the use of a module or shield. Some of the Teensy line of Arduino boards from pjrc.com as well as the Raspberry Pi Pico can also function as a USB host.

Figure 6.5: The SparkFun USB Host Shield.

A far simpler and less expensive method to add USB Host capability to your Arduino project is to use a USB Host shield (Figure 6.5) or module (Figure 6.6). Both are supported by the USBHost library and function identically. In actuality, the USB Mini Host Module (Figure 6.6) is a shield designed for the Arduino Pro Mini, but it works equally well as a module with other Arduino boards such as the UNO or Nano.

Figure 6.6: The USB Host Module/mini-shield.

Amazingly, through the use of the USB Host library, you can connect all manner of USB devices simply and easily. While I have not personally tried it yet, it should be feasible to also add Bluetooth HID capability through the use of a Bluetooth USB dongle plugged into the Arduino's USB host shield/module port. This is definitely at the top of the list of things to test, as it opens up a whole new world of simple and inexpensive wireless Arduino projects.

WiFi

The Arduino is rapidly evolving into a powerful Internet of Things (IoT) device. By using a WiFi-capable Arduino such as the ESP32 or the Arduino R4 WiFi, or by adding a WiFi shield to an Arduino such as the UNO, you can connect your Arduino projects to your local network and also the Internet itself. Note that the official Arduino UNO WiFi shield has been retired, however functionally equivalent shields are available from most of the online Arduino board suppliers, including Adafruit and SparkFun.

The CAN Bus

Figure 6.7: The CAN Bus shield.

The CAN (Controller Area Network) Bus is a bus standard used primarily in the automotive industry and designed to allow microcontrollers and vehicles to communicate. By using a CAN-BUS shield (Figure 6.7) or an Arduino with an onboard CAN bus interface such as the Arduino UNO R4 series, you can interface to and communicate with the industry standard CAN automotive diagnostic bus found on most modern cars and some newer machine tools. This allows you to access information from a vehicle's Electronic Control Units (ECUs) and read data such as throttle position, engine RPM, speed, and many other vehicle operating parameters.

Interrupts

While not often seen as an Arduino I/O method, interrupts allow external conditions and events to modify the way your Arduino sketches execute. The Arduino has two types of interrupts – hardware and timer. Hardware interrupts are triggered by an external event, such as a change in the logic level on a digital input pin. An interrupt will pause (interrupt) the current program execution and immediately execute a user-defined function, known as an interrupt handler or Interrupt Service Routine (ISR). When the ISR function is complete, program execution resumes the program normally with the next statement that follows the command that was last executed prior to the interrupt. An interrupt can happen at any time and allow your Arduino sketch to immediately respond to external events without having to constantly check (or poll) to see if the desired condition exists.

There are four types of interrupt conditions that can be defined: Rising, Falling, Change, and Low. These interrupt conditions refer to the state of the digital I/O pin used to generate the interrupt. The Rising condition will generate an interrupt when the I/O pin goes from a logic-level low state to a high state; Falling will generate an interrupt when the I/O pin goes from high to low. The Change condition will generate an interrupt when the I/O pin changes from either low to high or high to low. The Low condition will generate an interrupt when the I/O pin is low. The Arduino Due also has an additional interrupt condition, High, which generates an interrupt when the pin is high.

The Arduino UNO has two interrupts, assigned to digital I/O pins 2 and 3. The Arduino Leonardo has four interrupts, on Pins 0, 2, 3, and 7, and the Arduino Mega2560 has six interrupts, on pins 2 and 3, and 18 thru 21. The Arduino UNO can also handle Change interrupts on all I/O pins, but unlike hardware interrupts on the defined interrupt pins, the ISR function must decode the interrupt and determine which I/O pin generated the interrupt.

The Arduino UNO has three internal timers, defined as Timer0, Timer1, and Timer2, which can be used to generate Timer interrupts. Timer0 is an 8-bit timer used by the Arduino for internal timing functions such as delay() and millis(). Since it can affect these functions, modifying the Timer0 settings is not recommended. Timer1 is a 16-bit timer often used by some Arduino sketch libraries, such as the Servo library, and Timer2 is an 8-bit timer used by the Arduino tone() function. As long as you are aware of any potential interaction with these functions, you can modify the settings on Timer1 and Timer2 for use in your sketches. The Mega2560 has three additional 16-bit timers – Timer3, Timer4, and Timer5 – which are not used by any Arduino internal functions. Other Arduino and variant boards also have additional timers available onboard.

You can configure the timers to generate a software interrupt on overflow, or when the timer count reaches a desired value. The timers are based on the Arduino CPU clock rate (i.e., 16 MHz on the Arduino UNO). You can use the Timer Counter/Control Register (TCCR) for each timer to control the timer clock setting. By modifying the Clock Select bits, you can control how fast the timer increments the counter. On the Arduino UNO, the available settings are Clk/1 (clock speed), Clk/8, Clk/64, Clk/256, and Clk/1024. At the maximum setting of Clk/1024, you can have 16-bit Timer1 generate a software interrupt approximately every 4,194 seconds. Using the Clear Timer on Compare Match (CTC) setting, you can adjust the time to generate an interrupt when the timer reaches a preset value. For example, using 15624 as the preset value will cause the timer to generate an interrupt once per second. Using a 1-second interrupt in this manner, you can add precision timing to your sketches without having to manually keep track of time using the millis() function or other manual methods.

Implementing interrupts does add a level of complexity to your Arduino sketches, and you have to remind yourself that an interrupt can occur at any time during program execution. You will have to remember and plan for this as you write and troubleshoot your sketch. You can enable and disable the interrupts as needed from within your sketches, to allow for uninterrupted execution of critical or time-sensitive points in your sketch where you don't want the sketch execution to be interrupted.

Used properly, interrupts can be a powerful tool in developing your Arduino projects, and once you get comfortable using interrupts, you will find that they can greatly simply your project development, since you no longer need to have your sketches running in timing loops waiting for an event to occur. Instead, you can have your sketch off doing other things, and only respond when the actual event occurs.

You are not just limited to standalone projects with the Arduino. There are several applications, such as Processing, Ardulink, and others, which allow you to control your Arduino using its USB/Serial port from a workstation. With the addition of an inexpensive Bluetooth module, or by using an Arduino board that has built-in Bluetooth, you can control your Arduino from an Android smartphone or tablet.

Processing

The final I/O method isn't so much an I/O method as it is a means of combining the sense and control functions of the Arduino with the computing power of a workstation. Processing is an Open Source programming language created by Ben Fry and Casey Reas for the electronic arts and visual design community. Since the Arduino IDE is based on the Processing project, the Integrated Development Environment (IDE) for the two are very similar, which makes Processing an easy-to-learn choice for visualizing your Arduino data on a workstation.

Processing is primarily a design and prototyping tool used to create motion graphics and complex data visualization projects, including animation in both 2D and 3D. Processing was designed for graphic artists and designers rather than computer experts, so as with the Arduino, Processing is easy to learn and use. Processing is more of a graphics language than a traditional programming language, which makes it ideal for graphing and displaying trending-type data over time such as temperatures, voltages, etc.

As with the Arduino, programs written in Processing are also known as sketches, and your sketches are also located in a location known as the sketchbook. Processing can also use libraries, to add additional functionality and features to your Processing sketches such as networking, video, audio, animation , and much more. There's even a library to allow you to project a 3D Processing sketch on a spherical dome, like you would find in a planetarium.

The Arduino Firmata library allows you to use Processing to control the I/O pins on the Arduino without having to write any Arduino sketches. On the Arduino, all you have to do is upload the Firmata firmware instead of a sketch, and then you can use Processing to do the rest. This opens up some interesting possibilities in the way of Computer Aided Transceiver (CAT) control, with full graphic screen displays of the transceiver and operating controls.

An excellent book on learning how to use Processing is *The SparkFun Guide to Processing* by Derek Runberg (ISBN: 978-1-59327-612-6). There are also a number of tutorials online, including www.processing.org/tutorials, as well as an Open Source Processing sketch repository at: www.openprocessing.org.

ArduGraph

In addition to the Serial Plotter function embedded in the Arduino IDE, you can use the ArduGraph package. Created with Processing, ArduGraph allows you to create and display up to 3 simultaneous graphs from an Arduino. Easy to set up and configure, ArduGraph communicates with the Arduino over the USB/Serial port to plot graphs in real-time. Since ArduGraph uses the USB/Serial port, there is no need to install a library on the Arduino. All you have to do is add the ArduGraph commands to the Arduino's Serial.println() command and the data will be plotted on the selected graph. Since ArduGraph is written in Processing, you can convert ArduGraph to run as an executable file on a Windows, Linux, or Mac workstation using the Export Application feature on the Processing IDE.

Chapter 7 • Project 1 - CW Beacon and Foxhunt Keyer

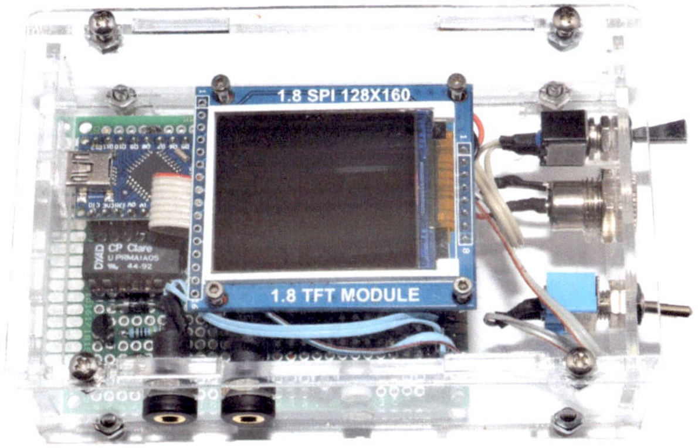

Figure 7.1: The finished CW Beacon and Foxhunt Keyer.

The inspiration for this project came from a friend and fellow Arduino builder, Tim Billingsley, KD5CKP (SK), who operated a 10-meter beacon in nearby Olive Branch, MS. In fact, it was Tim who introduced me to the Arduino in 2011.

Whenever 10 meters would open up, his beacon would get reports from all over the world with just three Watts of power out. Unfortunately, the beacon was finicky at best and had to be kicked back to life on a regular basis. At the same time, there was an upswing in interest among the local clubs to conduct Transmitter Hunts (Foxhunts). This sounded like a perfect opportunity to build a small CW Beacon and Foxhunt Keyer with the Arduino.

Starting out with the project flowchart (Figure 7.2), I originally designed a monster, first with a Nokia 5110 display module, then with a scrolling Organic LED (OLED) display module, and then added all sorts of bells and whistles. Soon, my design had expanded beyond the capacity of the small Arduino enclosure I had originally envisioned for it, so I had to go back to the drawing board and rethink exactly what I was trying to accomplish. In the end, with the consideration that this was to be a small, rugged, portable device, the design was simplified to be a simple beacon keyer to drive a keying relay and CW tone generator to create the modulated CW tones needed to key a 2-meter handheld for the Foxhunt mode. This redesigned version used simple LEDs to indicate the current status, however, in the finished prototype, the LEDs were replaced with a color TFT to provide more status information, while still maintaining the simplicity and low-power design of the project.

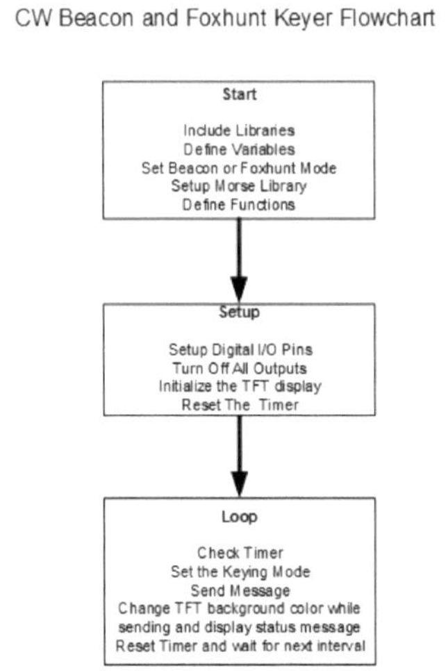

Figure 7.2: The CW Beacon and Foxhunt Keyer flowchart.

Keeping the concept as simple as possible, I next created the circuit block diagram in Figure 7.3. To display the status of the project, I decided to use a 160 x 128 ST7735-type color TFT display, allowing me to display the various states of the keying cycle and change the background color to give a visual status indication. Since the time intervals and beacon text would be switch-selectable, the values would be hard-coded within the sketch and could easily be changed as needed. A switch was added to allow the mode to be selected without requiring modifying the sketch.

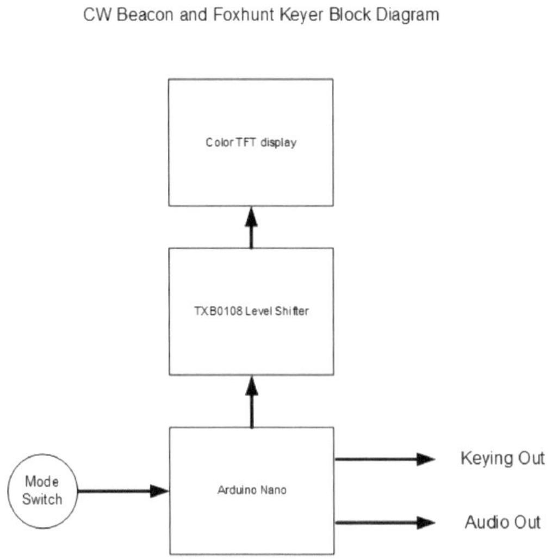

Figure 7.3: The CW Beacon and Foxhunt Keyer block diagram.

The schematic diagram for the CW Beacon and Foxhunt Keyer is shown in Figure 7.4 and the Parts List is shown in Figure 7.5. To keep the beacon keyer as versatile as possible, a small DIP reed relay driven by a 2N2222A transistor was used to isolate the Arduino keying signal from the transmitter. This type of circuit is a good idea anytime you want to drive a relay, as the Arduino digital I/O pins can only drive 40mA and some of the newer Arduino-type boards can only source 6 mA of current. Using a transistor to key a relay requires much less current and helps keep you from drawing too much current and damaging your digital I/O pin. A relay also isolates the Arduino I/O pin from your transmitter, keeping potentially damaging voltages at bay. The keying relay will be used to either act as a CW key for the Beacon transmitter or as the Push-To-Talk (PTT) keying on the Foxhunt transmitter. A mode switch was incorporated to allow switching between beacon and foxhunt modes.

A potentiometer was placed on the audio output to allow the output level to be adjusted to whatever signal level the Foxhunt transmitter would need. A level shifter is used between the Arduino Nano and the TFT display since the Arduino Nano is a 5 volt device and most TFT display modules are 3.3 volt devices. While some of the ST7735-type TFT displays are 5 volt-tolerant, why take chances and risk damaging the display module?

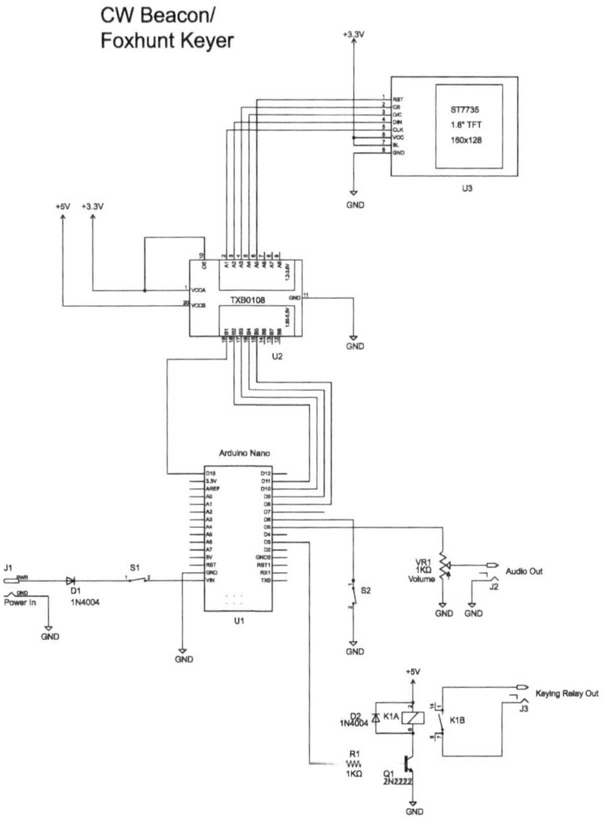

Figure 7.4: The CW Beacon and Foxhunt Keyer schematic.

Parts List
CW Beacon/Foxhunt Keyer

Part	**Value**
D1, D2	1N4004 Diode
J1	DC Power Jack
J2, J3	stereo mini phone jack
K1	PRMA1A05 (or similar) 5-V SPST DIP reed relay
Q1	2N2222 transistor
R1	1 kΩ resistor, 1/8W
S1, S2	SPST switch
U1	Arduino Nano
U2	TXB0108 8-channel Level Shifter module
U3	ST7735 1.8" color TFT display
VR1	1 kΩ potentiometer
Enclosure	Solarbotics Mega S.A.F.E

Figure 7.5: The CW Beacon and Foxhunt Keyer parts list.

Once I had laid out the circuit on my breadboard to match the schematic diagram, it was time to write the Arduino sketch. The entire CW Beacon and Foxhunt keyer sketch and the libraries for this project are available online at www.kw5gp.com/Elektor.

The CW Beacon portion of the sketch was to repeat the beacon message at a selected interval, while the Foxhunt section was to repeat the message for a selected interval, then go silent for different time interval. Since the Arduino does not have a Real Time Clock, there would need to be some way to keep track of time. Fortunately, for a simple project, high accuracy in the timing is not critical, so I was able to use a simple function to calculate time intervals using the Arduino millis() function.

At the heart of this project, all we are doing is converting text to a CW message, so we can use the Morse library to send the desired message. The Color TFT will be used to indicate the status. The configuration information will be displayed on the TFT and the background color will be used to indicate whether the keyer is idle or transmitting. A blue background would indicate that the keyer was operational and waiting to send a message as shown in Figures 7.6 and 7.7. A red background will be used to indicate that the keyer was sending a message as shown in Figure 7.8 and 7.9. The TFT text and background color selection will be handled by TFT library functions from within the sketch itself.

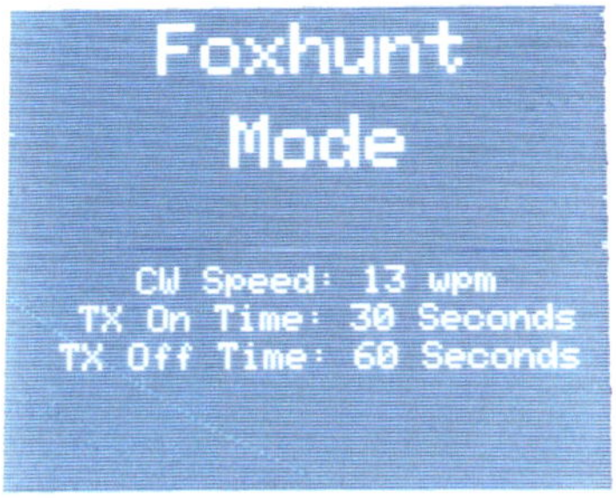

Figure 7.6: The CW Beacon waiting to transmit in Foxhunt Mode.

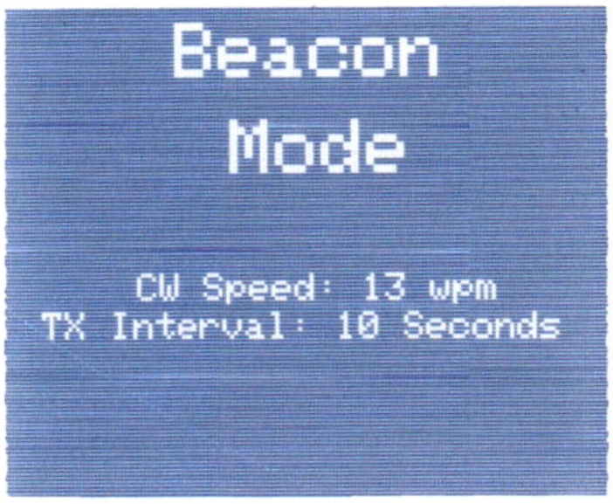

Figure 7.7: The CW Beacon waiting to transmit in Beacon Mode.

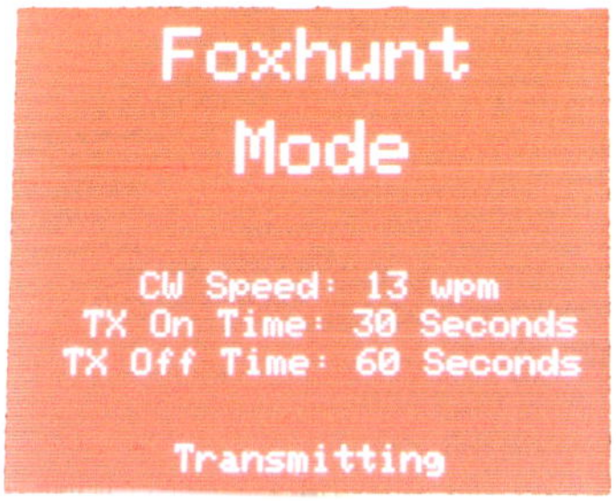

Figure 7.8: The CW Beacon transmitting in Foxhunt Mode.

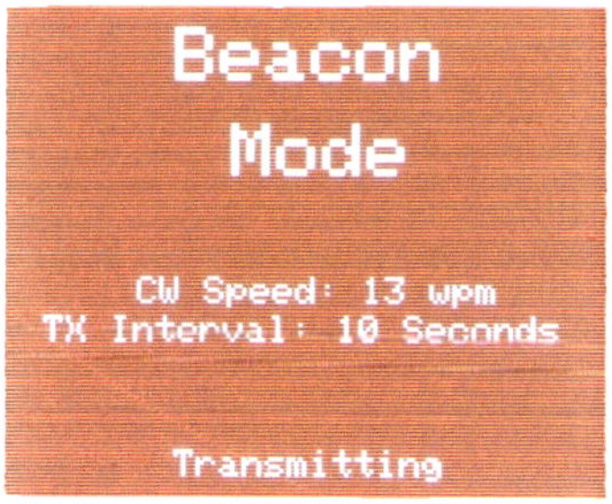

Figure 7.9: The CW Beacon transmitting in Beacon Mode.

As we go through the sketches for the projects, the entire sketch will not be printed here. Please download the sketches for this book from www.kw5gp.com/Elektor. As we start out with the sketch, to keep things simple and readable, we include the libraries needed for the sketch. Note that for this sketch to work properly, you must use the Morse library from the www.kw5gp.com/Elektor website as this library had minor errors that I had to correct.

```
#include <Morse.h
#include <Adafruit_GFX.h       // Core graphics library
#include <Adafruit_ST7735.h    // Hardware-specific library
```

Next, we define the constants that are used to configure the keyer. These defined values will be used to set the keyer to Beacon or Foxhunt mode, the beacon message, the size of the message array, the keying speed, the timer interval, and the number of seconds to repeat the message until the timer interval expires. We also define the digital I/O pins assigned to the keyer outputs. I prefer to define constants when I can, rather than declaring them as variables. This creates an IDE compiler pre-processor command to substitute the definition for the actual value as it compiles. This method saves valuable Static RAM (SRAM) that we may need for something else. Since constant values are just that, constant and unchanging, they can be placed in Flash memory through the use of the #define statement.

```
#define beacon_msg "DE KW5GP Beacon"
#define key_pin 3                  // Pin for CW tone
#define mode_pin 6                 // Pin for the Mode switch
#define beep_pin 5                 // Pin for PTT/Rig Key
#define key_speed 13               // CW send speed
#define beacon_interval 10         // Transmit every 10 seconds in foxhunt mode
#define foxhunt_interval 60        // Transmit every 60 seconds in foxhunt mode
#define TFT_CS     10              // Assign the TFT CS to pin 10
#define TFT_RST    8               // Assign the TFT RST to pin 7
#define TFT_DC     9               // Assign the TFT DC to pin 6

#define tft_delay 10               // set the TFT command delay to 10ms
```

The timer() function will return the time in seconds since the timer was reset by the TimerReset() function. This is how the keyer keeps track of the various time intervals. When the TimerReset() function is called, it doesn't actually reset anything. All it does is set the offset variable to the current millis() value, which equates to the current time in milliseconds since the Arduino was last reset. We then use this offset value to calculate our timing intervals.

```
// Timer function - returns the time in seconds since last Timer Reset
unsigned long Timer()
{
  return (millis()- offset)/1000;  // return time in seconds
}

// Timer Reset function
```

```
void TimerReset(unsigned long val = 0L)
{
  offset = millis() - val;
}
```

In the setup() loop, the TFT is initialized. Please not that there are multiple versions of the 1.8" color ST7735-type TFT displays, typically indicated by a colored tab on the TFT screen's clear protective cover. You may need to change the tft.initR command to match the color of the tab on your TFT display module. The correct initialization keyword for other color tabs can be found in the example sketches for the Adafruit ST7735 library.

```
  tft.initR(INITR_18BLACKTAB);   // initialize a 1.8" TFT with ST7735S chip,
black tab
```

As part of the TFT setup, I like to display a "Splash" screen to indicate that the display is working and to let me know what sketch is loaded on the Arduino. Notice that there is a small delay (defined as 10 millliseconds) between TFT display commands. Some of the generic TFT displays do not operate as fast as other displays of the same type, causing the display to not operate correctly. Adding this small delay between commands will usually resolve this issue.

```
  delay(tft_delay);
  clear_display();              // Clear the display - fill with BLACK background
  delay(tft_delay);
  tft.setRotation(1);           // Set the screen rotation
  delay(tft_delay);
  tft.setTextWrap(false);       // Turn off Text Wrap
  delay(tft_delay);
  tft.setTextSize(3);           // Set the Font Size
  delay(tft_delay);
  tft.setTextColor(ST7735_WHITE); //Set the Text Color
  delay(tft_delay);
  tft.setCursor(40, 10);        //Set the Cursor and display the startup screen
  delay(tft_delay);
  tft.print("KW5GP");
  delay(tft_delay);
  tft.setTextSize(2);
  delay(tft_delay);
  tft.setCursor(40, 60);
  delay(tft_delay);
  tft.print("Foxhunt");
  delay(tft_delay);
  tft.setCursor(65, 80);
  delay(tft_delay);
  tft.print("and");
  delay(tft_delay);
```

```
tft.setCursor(10, 100);
delay(tft_delay);
tft.print("Beacon Keyer");
delay(tft_delay);
```

After setting up the TFT display, the digital I/O pin modes are set up and the keyer is ready to begin operating.

```
pinMode(key_pin, OUTPUT);               // set PTT/Key pin to output
pinMode(mode_pin, INPUT_PULLUP);        // Set the Mode pin for input with pullup
digitalWrite(key_pin, LOW);             // Turn off the PTT/Key relay
```

In the main loop(), the sketch sets the mode of operation. In Beacon Mode, the keyer will act as a CW key on the defined I/O pin, with the relay functioning as the key to send a CW message at the defined speed. In Foxhunt mode, the keyer will send an audio tone used to transmit a tone on the transmitter, on the defined I/O and at the defined speed. The relay I/O pin is then used to key the Foxhunt transmitter when in Foxhunt Mode. The transmitting interval is also set to the pre-defined interval for the selected mode.

If we're in Foxhunt mode, the keyer will continue to send the message until the repeat interval expires and will also add a space in between messages. In Beacon mode, the message is sent once.

When it's time to transmit, the display background will be changed to red and if we're in Foxhunt mode, the relay for the PTT line will be energized.

```
ptt = HIGH;
digitalWrite(key_pin, HIGH);  // Key the PTT
update_display();

Next we'll send the actual message:

while (Timer() < end_time) // Check to make sure repeat timer has not expired
(Foxhunt Mode)
    {

      for (int x = 0; x < msg_length; x++)  // Send the message in the beacon_
call character array one character at a time
      {
        c = beacon_call[x];  // Get the next letter to send
        morse.send(c);  // Send it in CW
      }

      if (mode == 1)  // Send a space if we're in Foxhunt mode to separate
messages
      {
```

```
      morse.send(char(32));
      delay(1000);

    }
    else
    {
      end_time = Timer() - 1;
    }
  }
```

After the message is sent, the timer is reset, and the keyer will change the TFT display background back to blue to indicate that the keyer is idle and waiting for the next transmit cycle. We'll also check to see if the mode switch has been changed and set the appropriate mode if the switch position has changed.

As I construct a project, I'll update the schematic diagram to address any design errors or changes. When the sketch and prototype are completed and working, I'll use this updated schematic as a guide to move the electronic components to a 60 cm x 80 cm copper-clad prototyping board as shown in Figures 7.10 and 7.11 and mount the finished project into a Solarbotics MegaSAFE Arduino Project Enclosure as shown at the beginning of this chapter. The Solarbotics enclosure is small enough to be mounted to the back of a handheld transceiver, allowing it to be used as a self-contained Foxhunt keyer.

Figure 7.10: The top of the CW Beacon and Foxhunt Keyer circuit board.

Figure 7.11: The bottom of the CW Beacon and Foxhunt Keyer circuit board.

Enhancement Ideas

There are a number of enhancements you can add to this project. Originally, as part of the Foxhunt keyer, I had planned to add a pushbutton that could be pressed to indicate that the transmitter had been found and the beacon message would be changed to a different message. You could also use a small very lower power Organic LED (OLED) display to scroll the actual beacon message as it is being sent and to indicate when the "found" button was pressed.

The last enhancement I would add is probably the main reason they don't put me in charge of the fox in Foxhunts anymore. Since the Arduino is pretty good at robotics too, I'd love to mount the transmitter and Foxhunt keyer on top of an Arduino-powered robot and have it move 50 feet in random directions in between transmissions, install proximity sensors to avoid obstacles, and have motion detectors to sense when someone is nearby and have it move away from them or hunker down and go silent. Now that's a real Foxhunt.

Chapter 8 • Project 2 – Mini Weather Station

Figure 8.1: The finished Mini Weather Station.

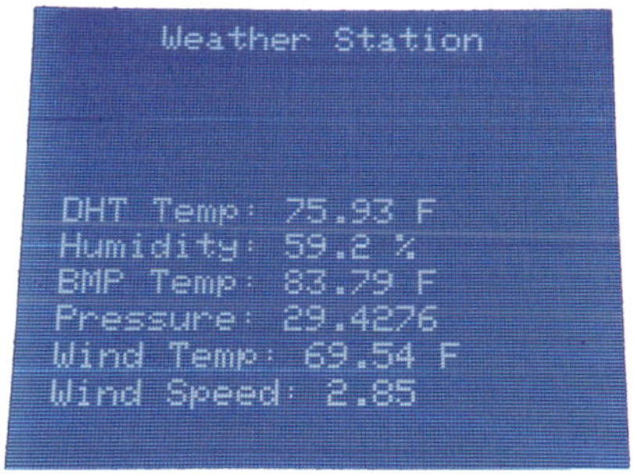

Figure 8.2 The finished Mini Weather Station display.

The day I began creating this project, my weather radio was sounding all sorts of weather alerts. Almost constant Severe Thunderstorm, Tornado Watch, Tornado Warnings, and Flash Flood warnings had been going off for most of the day, and the local ham emergency nets were all active. Perfect timing for a Weather Station project wouldn't you say? Hams and weather seem to go hand in hand, from storm chasers, to weather emergency groups and nets, to just plain old weather watching, many hams have an interest in the weather.

The Arduino has a wide array of sensors that can be used to monitor many different weather and atmospheric conditions. When I was growing up, our family had one of the wall-mounted weather stations that displayed the temperature, relative humidity, and barometric pressure on analog dials. For this project, I decided to revisit my childhood memories and give that wonderful old weather station an Arduino-powered makeover.

Moving forward in our Arduino adventures, the Mini Weather Station shown in Figures 8.1 and 8.2 uses an ASAIR DHT20 Temperature and Relative Humidity Sensor, a Bosch BMP085 Barometric Pressure Sensor, and a very interesting new module, a Wind Sensor from Modern Device. This project also uses the ST7735-type TFT display module, which has become my display of choice for many Arduino projects. All of these devices, with the exception of the Wind Speed Sensor, are well supported with Arduino libraries and example code. The good news is that we can get what we need for the Wind Speed Sensor from the example files on the manufacturer's website.

Figure 8.3: The ASAIR DHT20 relative humidity and temperature sensor.

The ASAIR DHT20 shown in Figure 8.3 is an upgraded version of the venerable DHT11 Temperature and Humidity sensor that communicates with the Arduino via the I²C bus. The Relative Humidity portion of the DHT20 has a resolution of 0.024% humidity and an accuracy of ±3% over a temperature range of –40 °C to 80 °C (–40°F to 176 °F). The Temperature sensor in the DHT20 has a resolution of 0.01 °C with an accuracy of ± 0.5 °C over the same operating range, far more accurate and with a wider range than the DS18B20 Temperature sensor often used to sense temperature with the Arduino.

Figure 8.4: The BMP280 Pressure Sensor module.

The Bosch BMP280 Pressure sensor shown in Figure 8.4 communicates with the Arduino using the I^2C bus. The companion BME280 sensor adds Relative Humidity capability to the sensor. Theoretically, you could replace the DHT20 Relative Humidity and Temperature module and the BMP280 Pressure modules with a single BME280 module to sense all three environmental conditions. I chose to use separate sensor modules as a learning experience for using multiple modules on the I^2C bus.

The BMP280 has an absolute accuracy of ±1 hPa over an operating temperature range of –40° to +85°C. As with the DHT20, the BMP280 also has an embedded temperature sensor that has a resolution of 0.01°C with an accuracy of ±1°C over the same operating temperature range as the barometric pressure sensor portion of the module. The BMP280 sensor module requires 3.3 volts to operate, however many of the more recent modules have onboard 5 Volt regulators. This project uses the 5 volt version of the BMP280 module.

Pascals, Millibars and Inches of Mercury

When I began working with the Mini Weather Station project, I quickly discovered that the Bosch BMP280 Barometric Pressure sensor outputs its data in hectoPascals. Hecto what? Maybe I've led a sheltered life, but up to this point I have always known Barometric Pressure to be in millibars or inches of mercury. Some quick research on the web provided the formulas and information needed to convert these hectoPascal things to numbers I was used to.

A Pascal (Pa) is equal to 100 millibars, or, using the hectoPascals (hPa) output from the BMP085, 1 hPa = 1 millibar of pressure. One inch of mercury (inHg) is equal to 3386.489 Pascals (33864.89 hPa) at 0 °C. For reference, one standard atmosphere at sea-level is defined as 29.92 inches of mercury, or 1013.25 hPa.

Since the standard pressure at sea-level can be used as a reference point, we can apply the formula in the BMP280 library and example sketch to our barometric pressure sensor data and calculate our approximate altitude above sea-level. So, not only can you use your barometric sensor to measure air pressure, with just a library function you can turn it into an altimeter as well.

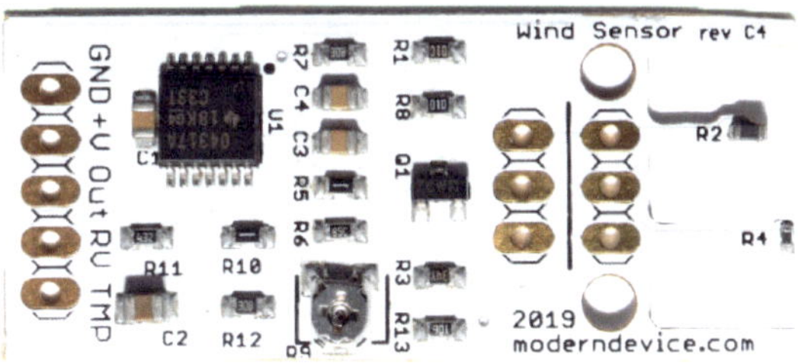

Figure 8.5: The Modern Device Wind Speed Sensor module.

A recent addition to the Arduino environmental sensor lineup is the Wind Speed Sensor module from Modern Device. The interesting part about this module is that it measures wind speed without any moving parts as you would find in a typical "spinning-cup"-type anemometer. Instead, it uses a measurement technique known as "hot-wire". The hot-wire technique involves heating an element to a precise constant temperature and then measuring the amount of power needed to maintain that temperature as the wind speed changes. In addition to measuring low to medium wind speeds, the hot-wire technique is an excellent way to measure indoor air movement where a standard spinning-cup anemometer would be too bulky and simply inefficient.

Some interesting uses for this wind speed sensor module mentioned on the Modern Device website include human breath detection, room occupancy detection, HVAC system monitoring, and of course, weather stations. Having recently had HVAC system airflow issues, an Arduino-powered HVAC monitor using this sensor is definitely on my list of Arduino projects to build.

Starting out with the usual block diagram (Figure 8.6), I wanted the Arduino to read the DHT20, BMP280, and Wind Speed sensor and display the current temperature, relative humidity, barometric pressure, and wind speed on the 1.8" ST7735-type Color TFT display.

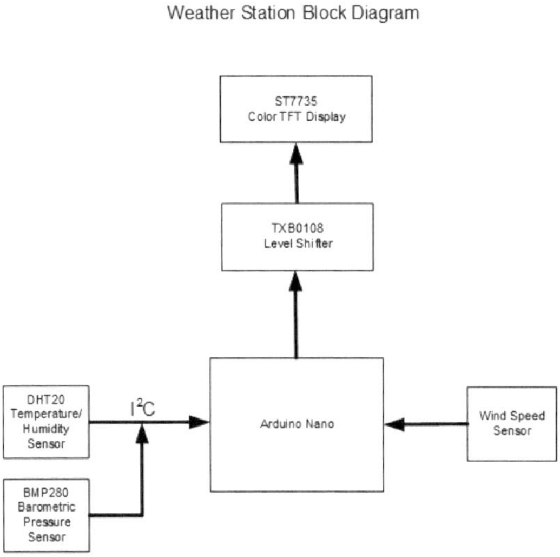

Figure 8.6: Mini Weather Station block diagram.

Using the block diagram as a guide, the next order of business was to create a draft schematic and wire the project up on the breadboard. Using the schematic diagram I created for this project (Figure 8.7) and the Parts List (Figure 8.8), I wired up the project on my breadboard. You will notice that there are no pull-up resistors on the I²C bus for the BMP280 and DHT20 sensors. The BMP280 module used in this project is the 5 volt version and it comes with 10KΩ I²C bus pull-up resistors already mounted on the module itself. Because of this, no additional pull-up resistors are needed for the DHT20 module and therefore, the entire I²C bus. This is something important to note. Many I²C modules have pull-up resistors on the board. Check to see if any module you're installing on the I²C bus already has pull-up resistors before adding any additional pull-up resistors.

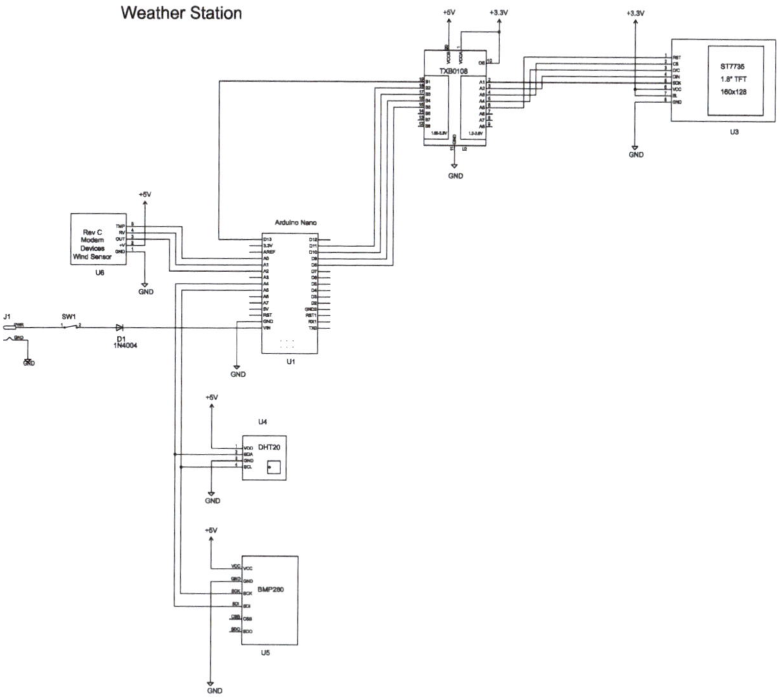

Figure 8.7: The Mini Weather Station schematic.

Parts List
Weather Station

Part	**Value**
D1	1N4004 diode
J1	DC power jack
SW1	SPST switch
U1	Arduino Nano
U2	TXB0108 8-channel level shifter module
U3	ST7735 1.8" color TFT display
U4	DHT20 humidity and temperature sensor
U5	BMP280 barometric pressure and temperature sensor
U6	Modern Devices Wind Speed Sensor Rev. C
Enclosure	Solarbotics MegaSAFE

Figure 8.8: The Mini Weather Station parts list.

While planning out the sketch flowchart (Figure 8.9), I found a number of libraries available for the DHT20, BMP280, and the ST7735-type TFT display. The libraries I chose to use were the DFRobot DHT20 library, the Adafruit BMP280 library, and the Adafruit ST7735 and Adafruit GFX libraries, although any of the other libraries would do just as well. At the time this project was created, I could not locate a library for the Modern Design Wind Speed Sensor, so the sketch for the Wind Speed Sensor does not use a library, and the Modern Design Example sketch from their website was used as a guideline instead. Most of the complex operations are handled by the libraries or by the functions in the example sketches, which greatly simplifies the use of these sensors in your projects.

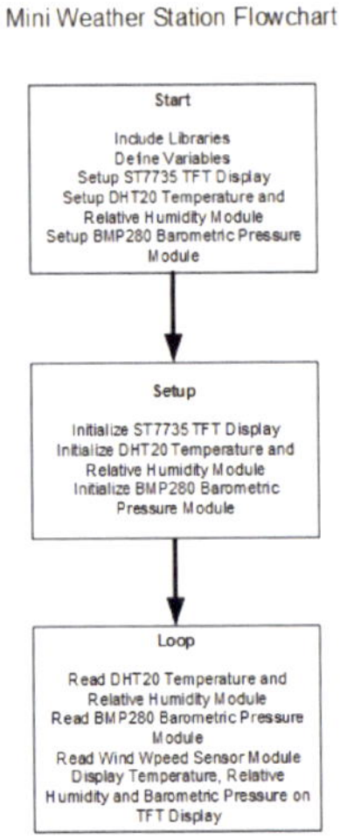

Figure 8.9: Mini Weather Station flowchart.

As you can tell by looking at the sketch, there is a lot of initialization needed to get the sensors and display ready for operation. Because we are using I²C for the DHT20 and BMP280, we also must include the Arduino internal I²C Wire.h library. Often, you will not need to include the Wire.h library as it has been included inside the device's library include statements. I always add it just in case, as there's no harm in including a library twice since the IDE will ignore duplicate include statements when it compiles and uploads the sketch. The complete sketch and required libraries for this project can be found at www.kw5gp.com/Elektor.

To start in the sketch, we'll include the libraries and define the I/O pins we'll need for the various modules.

```
#include <Wire.h> // I2C Library
#include <Adafruit_BMP280.h>   //Adafruit BMP280 Library
#include <Adafruit_GFX.h>      // Core graphics library
#include <Adafruit_ST7735.h>   // Hardware-specific library
#include <DFRobot_DHT20.h>

#define analogPinForRV    1    // The Wind Sensor RV Pin
#define analogPinForTMP   0    // The Wind Sensor TMP pin

#define TFT_CS     10          // Assign the TFT CS to pin 10
#define TFT_RST    8           // Assign the TFT RST to pin 7
#define TFT_DC     9           // Assign the TFT DC to pin 6
```

Next, the variables for the Wind Speed sensor, the DHT20, and the BMP280 need to be declared. The zeroWindAdjustment constant comes directly from the Wind Speed Sensor example sketch and could probably be changed to a #define pre-processor statement, but for this sketch I chose to use the original example variable declarations.

```
const float zeroWindAdjustment =  .2; // negative numbers yield smaller wind
speeds and vice versa
int TMP_Therm_ADunits;                  //temp termistor value from wind sensor
float RV_Wind_ADunits;                  //RV output from wind sensor
float RV_Wind_Volts;
float TempCtimes100;
float TempF;
float zeroWind_ADunits;
float zeroWind_volts;
float WindSpeed_MPH;
unsigned bmp_status;
float bmp_temp;
float bmp_pressure;
float bmp_inches;
float dht_temp;
float dht_humidity;
```

Next, we'll instantiate the ST7735-type TFT Display, the BMP280 Pressure Sensor, and the DHT20 Relative Humidity and Temperature Sensor:

```
// Instantiate the TFT Display
Adafruit_ST7735 tft = Adafruit_ST7735(TFT_CS,  TFT_DC, TFT_RST);  // Initialize
the TFT display instance

// Instantiate the BMP280 sensor
Adafruit_BMP280 bmp; // use BMP280 I2C interface

// Instantiate the DHT20 sensor
DFRobot_DHT20 dht20;
```

In the setup() loop, the DHT20 module is started, then the TFT display is initialized and a startup message is briefly displayed before starting the BMP280 module.

```
  dht20.begin(); // Start the DHT20 sensor

  // Display the splash screen
  tft.initR(INITR_18BLACKTAB);   // initialize a 1.8" TFT with ST7735S chip,
black tab
  delay(tft_delay);
  clear_display(); // Clear the display - fill with BLACK background
  delay(tft_delay);
```

```
  tft.setRotation(1);          // Set the screen rotation
  delay(tft_delay);
  tft.setTextWrap(false);      // Turn off Text Wrap
  delay(tft_delay);
  tft.setTextSize(3);          // Set the Font Size
  delay(tft_delay);
  tft.setTextColor(ST7735_WHITE); //Set the Text Color
  delay(tft_delay);
  tft.setCursor(40, 10);       //Set the Cursor and display the startup screen
  delay(tft_delay);
  tft.print("KW5GP");
  delay(tft_delay);
  tft.setTextSize(2);
  delay(tft_delay);
  tft.setCursor(40, 60);
  delay(tft_delay);
  tft.print("Arduino");
  delay(tft_delay);
  tft.setCursor(40, 80);
  delay(tft_delay);
  tft.print("Weather");
  delay(tft_delay);
  tft.setCursor(40, 100);
  delay(tft_delay);
  tft.print("Station");
  delay(tft_delay);

  // Start the BMP280 - You may need to use the alternate address version with
  // non-Adafruit BMP Modules as I had to here with a generic BMP280 module

  //  status = bmp.begin();
  bmp_status = bmp.begin(BMP280_ADDRESS_ALT, BMP280_CHIPID);
```

In the main loop(), functions are used to read the BMP280 Pressure Sensor, the Wind Speed Sensor, and the DHT20 Relative Humidity and Temperature Sensor. The update_display() function is then used to update the ST7735-type color TFT display. By using function calls to perform the majority of the work, as you can see, the sketch main loop() itself consists of only four statements.

```
void loop()             // Main Program Loop
{
  get_bmp_data();       // Read the BMP280 and store the data into variables
  get_wind_data();      // Read the Wind Speed sensor and store the data into
variables
  get_dht_data();       // Read the DHT20 sensor and store the data into variables
  update_display();     // Update the display with the new data
```

```
  delay(update_delay);
}
```

The update_display() function is one that I commonly use in sketches that use a display module. This helps keep all of the display operations in one function, allowing for quick and easy troubleshooting and modification:

```
void update_display()  // Updates the mode on the display
{
// Display the data
  clear_display();
  delay(tft_delay);

  tft.setCursor(35, 5);
  delay(tft_delay);
  tft.print("Weather Station");
  delay(tft_delay);
  tft.setCursor(10, 55);
  delay(tft_delay);
  tft.print("DHT Temp: ");
  delay(tft_delay);
  tft.print(dht_temp, 2);
  delay(tft_delay);
  tft.print(" F");
  delay(tft_delay);
  tft.setCursor(10, 65);
  delay(tft_delay);
  tft.print("Humidity: ");
  delay(tft_delay);
  tft.print(dht_humidity, 1);
  delay(tft_delay);
  tft.print(" %");
  tft.setCursor(10, 75);
  delay(tft_delay);
  tft.print("BMP Temp: ");
  delay(tft_delay);
  tft.print(bmp_temp, 2);
  delay(tft_delay);
  tft.print(" F");
  delay(tft_delay);
  tft.setCursor(10, 85);
  delay(tft_delay);
  tft.print("Pressure: ");
  delay(tft_delay);
  tft.print(bmp_inches, 4);
```

```
    delay(tft_delay);
    tft.setCursor(10, 95);
    delay(tft_delay);
    tft.print("Wind Temp: ");
    delay(tft_delay);
    tft.print(TempF, 2);
    delay(tft_delay);
    tft.print(" F");
    delay(tft_delay);
    tft.setCursor(10, 105);
    delay(tft_delay);
    tft.print("Wind Speed: ");
    delay(tft_delay);
    tft.print(WindSpeed_MPH);
    delay(tft_delay);
}
```

Similarly, the methods used to read the DHT20, BMP280, and Wind Speed Sensor are all separated into individual functions. As you continue to work with the Arduino, you'll discover that breaking out as much of your sketch as you can into functions as I have done in this sketch, it will be much easier to re-use these functions in future project sketches.

```
void get_wind_data()  // Reads the data from the Wind Speed sensor
{
  // This portion is duplicated from the example sketch
  TMP_Therm_ADunits = analogRead(analogPinForTMP);
  RV_Wind_ADunits = analogRead(analogPinForRV);
  RV_Wind_Volts = (RV_Wind_ADunits *  0.0048828125);
  TempCtimes100 = (0.005 * ((float)TMP_Therm_ADunits * (float)TMP_Therm_ADunits))
- (16.862 * (float)TMP_Therm_ADunits) + 9075.4;
  TempF = ((TempCtimes100 * 9 / 5) + 3200) / 100;

  zeroWind_ADunits = -0.0006 * ((float)TMP_Therm_ADunits * (float)TMP_Therm_
ADunits) + 1.0727 * (float)TMP_Therm_ADunits + 47.172; //  13.0C  553  482.39

  zeroWind_volts = (zeroWind_ADunits * 0.0048828125) - zeroWindAdjustment;

  // This from a regression from data in the form of
  // Vraw = V0 + b * WindSpeed ^ c
  // V0 is zero wind at a particular temperature
  // The constants b and c were determined by some Excel wrangling with the
solver.

  //We can't have a negative number in the following calculation otherwise we get
a "nan" on the display
  // You can also adjust the zeroWindAdjustment constant to fine tune the
```

```
zeroWind_volts

  if (RV_Wind_Volts - zeroWind_volts < 0)
  {
    zeroWind_volts = RV_Wind_Volts;
  }
  else
  {
    WindSpeed_MPH =  pow(((RV_Wind_Volts - zeroWind_volts) / .2300) , 2.7265);
  }
}

void get_bmp_data() // Reads the BMP280 sensor
{
  bmp_temp = bmp.readTemperature();
  // convert temp to Farenheit
  bmp_temp = (bmp_temp * 9 / 5) + 32;

  bmp_pressure = bmp.readPressure() / 100;
  bmp_inches = bmp_pressure * .02953; // convert hPa to inches of Mercury
}

void get_dht_data() // Reads the DHT20 sensor
{
  dht_temp = dht20.getTemperature();
  // convert temp to Farenheit
  dht_temp = (dht_temp * 9 / 5) + 32;
  dht_humidity = dht20.getHumidity() * 100;
}
```

Construction Notes

With the sketch completed and the project working as planned on the breadboard, the schematic diagram for the project was used to construct everything on a 60 mm by 80 mm prototyping board as shown in Figure 8.10. The finished project was then mounted into a clear Solarbotics MegaSAFE enclosure as shown in Figure 8.11. I prefer to use the MEGA size enclosure as it allows more room for mounting external components such as a battery, switches, etc. To get more accurate wind speed, temperature, and humidity readings, the Wind Speed Sensor was mounted on a mast made from a small piece of balsa wood from Hobby Lobby. The DHT20 was mounted on the mast as well, in order to avoid heat from the circuit board affecting the readings.

I have also found the Solarbotics enclosures to be ideal for projects that use color TFT displays, as the display can actually be mounted inside the clear top cover of the Solarbotics enclosure using 2mm hardware. This allows the display to be clearly seen, while keeping it safe inside the enclosure.

Figure 8.10: The finished Mini Weather Station prototyping board.

Figure 8.11: The finished Mini Weather Station in the Solarbotics enclosure.

Enhancement Ideas

This is one of those projects where it can be difficult to choose what you want it to be as you have so many atmospheric and weather-related sensor options available. As soon as I finished building this project, I discovered the Rain and Water Sensor shown in Figure 8.12. This would be a great addition to any weather station, allowing you to track rainfall on top of the other environmental data.

Figure 8.12: The Rain and Water Sensor module.

Also, since much weather-related data involves a trend over a period time, adding a Real-Time Clock Calendar module and datalogging function and graphic trends using the graphic display capabilities of the color TFT display would be ideal. To further utilize the graphic capabilities of the TFT display, you can track the barometric pressure and add an up or down arrow next to the barometric pressure reading on the display to indicate that the pressure is rising or falling. Finally, you could use a text to speech module and have the weather station speak the time, temperature, etc. for you.

This project turned out to be a lot of fun for me and it will come in real handy in just a few short weeks when my local club does our annual exercise in insanity as we compete in the outdoor category of the annual Winter Field Day Association's Winter Field Day event (www.winterfieldday.org), where temperature is part of the contest exchange. Hopefully at some point, I'll get the Ethernet side of this project worked out and we can have the contest software automatically handle the time and temperature side of things for us as well.

Chapter 9 • Project 3 — RF Probe with LED Bar Graph

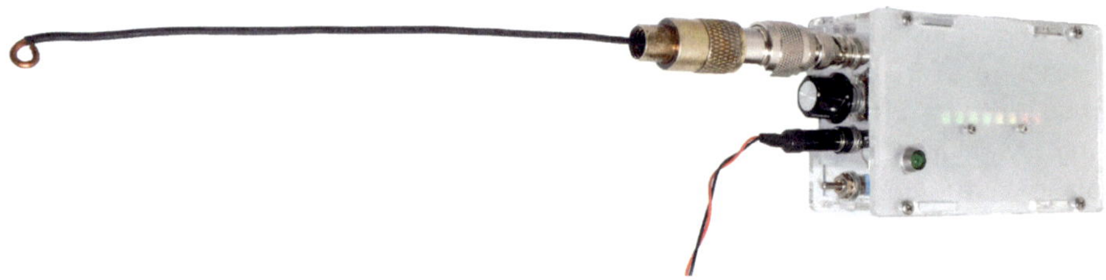

Figure 9.1:The finished RF Probe.

One of the cool things about the Arduino is that it can be used as a tool to make other tools. If you're like most hams, you don't often have a need for a lot of test equipment beyond a multimeter and a power/SWR meter, and when you do need to test something with a piece of equipment you don't have, you can usually borrow what you need from a fellow ham. The main reason for not owning your own cache of test equipment is primarily because some test equipment can be rather expensive, or it's just not worth the investment for something you'll only use once or twice in a blue moon.

In my case, I don't often need an RF probe, but it can come in handy when you want to know if that transmitter is actually transmitting and you don't happen to have a wattmeter or SWR meter handy, or maybe you need to track down some stray RF in the shack. While researching another project, I came across the *RF Driven On-Air Indicator* article by Keith Austermiller, KB9STR, in the August 2004 issue of *QST Magazine*, which itself is derived from the *"No Fibbin" RF Field Strength Meter* by John D. Noakes, VE7NI, in the August 2002 issue of *QST*. With a few minor tweaks, those projects could be adapted into an RF probe that would allow an Arduino to drive a WS2812-type addressable LED stick instead of a meter to indicate the relative strength of the RF signal.

Even though this appears to be a relatively simple and straightforward project, I still like to do the full documentation work-up. Having this documentation on hand can really help if I want to revisit this project at some later date and redesign it, add features, etc. Figure 9.2 shows the block diagram for the RF Probe project. In addition to the RF sensing input, the Arduino would be used to display the signal strength on an 8-LED WS2812-type RGB LED display stick.

The WS2812-type RGB LED stick is what really simplifies this project. Rather than having to drive discrete RGB LEDs at a cost of 3 digital I/O pins each and a total of 24 I/O pins for an 8-LED display or use 3 colors of LED's at a cost of one digital I/O pin each for a total of 8 I/O pins for an 8-LED display, the WS2812-type addressable RGB LED sticks and strips only need a single digital I/O pin. Alternatively, you could use a bar graph LED Driver chip such as the MAX7219 that would only require one digital I/O pin, but why add the complexity of a 24-pin chip when you don't have to?

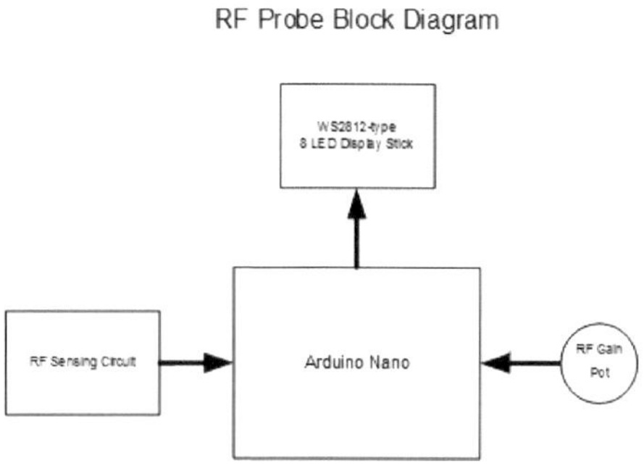

Figure 9.2: The RF Probe block diagram.

The construction of the RF Probe goes a little bit differently than the previous projects. Rather than prototype the RF Sensing unit on the breadboard and have to be concerned about the effects of RF on the breadboard wiring, the RF Sensing circuit is built and mounted on the same prototyping board as the Arduino Nano. Rather than use the Op Amp design used in the *RF Driven On-Air Indicator* article, I decided to modernize the circuit a bit and an AD8307 Logarithmic Amplifier was used instead, as shown in the RF Probe schematic diagram in Figure 9.3. The AD8307 can sense RF up to 500 MHz, so the RF probe project should be able to be used for all of your RF detection needs short of cell phone and WiFi frequencies. Since the output from the AD8307 is an analog voltage representing the signal strength, we can connect the output of the AD8307 directly to the Analog Input 0 pin of the Arduino Nano.

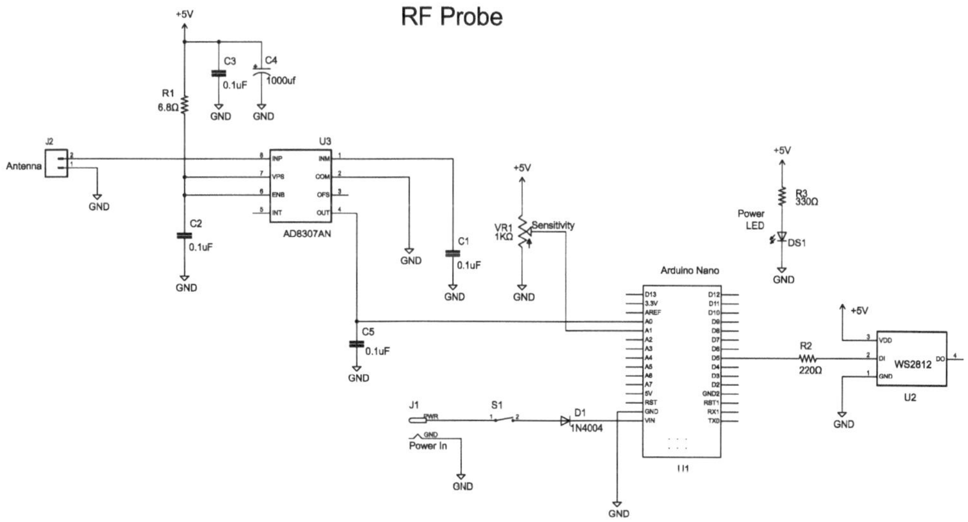

Figure 9.3: The RF Probe schematic.

Parts List
RF Probe

Part	**Value**
C1, C2, C3, C5	0.1 µF, 25 V capacitor
C4	1000 µF, 25V capacitor
D1	1N4004 diode
DS1	5 mm LED
J1	DC power jack
R1	6.8 Ω resistor, 1/8W
R2	220 Ω resistor, 1/8W
R3	330 Ω resistor, 1/8W
S1	SPST switch
U1	Arduino Nano
U2	WS2812 8-LED stick
U3	AD8307 logarithmic amplifier IC
VR1	1 kΩ potentiometer
Enclosure	Solarbotics Mega S.A.F.E.

Figure 9.4: The RF Probe parts list.

The antenna connector is mounted to a side panel of the Solarbotics enclosure and connects to the RF Probe circuit board using 0.1-inch (2.54 mm) male headers and the DuPont-style female pins inside connector shells to allow for easy removal of the circuit board to correct the inevitable wiring error. The AD8307 Logarithmic Amplifier chip is mounted in a socket for those times when you accidentally feed it 100 watts and let the smoke out. The RF probe is built on a 60 mm by 80 mm piece prototyping board as shown in Figure 9.5 and mounted into a clear Solarbotics MegaSAFE enclosure. An SO-239 coax chassis connector was mounted to the Solarbotics enclosure allowing you to use different antennas for the RF pickup antenna. A short 10-inch piece of #14 AWG solid wire was soldered to the center conductor of a PL-259 coax connecter to serve as the antenna for the RF sensing circuit.

Figure 9.5: The RF Probe circuit board.

For this project, we will be using the Arduino Nano's built-in 10-bit Analog to Digital Converter (ADC) to convert the analog voltage output from the AD8307 into a numeric value that is then converted to a number to indicate the number of LEDs to light that will serve as a representation of the signal strength. We'll then light the specified LEDs on an 8-LED WS2812-type RGB LED strip. For an RF Probe, you don't necessarily need the horsepower and graphics of a TFT display, as all you're really interested in is a relative RF Field Strength indication, which is perfectly suited to the WS2812-type LED strips. A 1 kΩ potentiometer will be used to adjust the RF sensitivity of the probe.

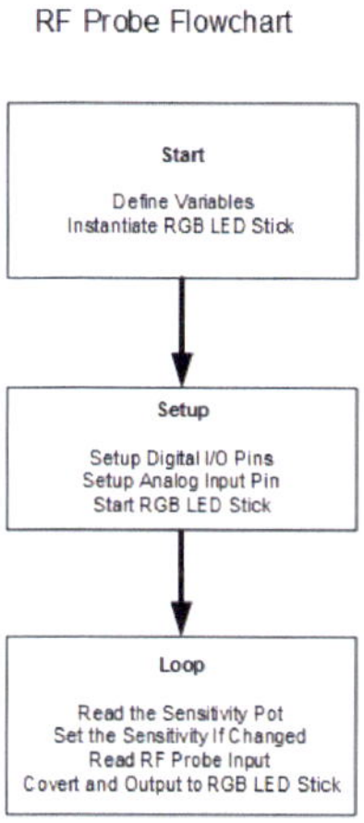

Figure 9.6: The RF Probe flowchart.

The sketch for the RF Probe is actually rather simple. Using the project Flowchart (Figure 9.6), the sketch itself is quite small, showing how well adapted the Arduino is to the simple I/O tasks used in this project. All the sketch has to do is read the analog voltage from the AD8307 and output a bar graph representation of that input voltage on the LED stick. The sketch will use 4 four LEDs to show green, 2 to show yellow, and the final 2 to show red, based on the strength of the signal detected as shown in Figure 9.7. You'll notice that I used a small piece of tissue paper glued to the inside of the enclosure's clear lid as a diffuser. You could also use one of the translucent Solarbotics enclosures as well.

Figure 9.7: The RF Probe enclosure.

We start the sketch for the RF Probe by including the Adafruit_NeoPixel library to drive the addressable 8 RGB LED stick. There are a number of libraries for the WS2812-type addressable RGB LEDs to choose from, but I prefer all of the features and flexibility in the Adafruit library.

```
#include <Adafruit_NeoPixel.h>
```

Next, we'll define the number of LEDs in the addressable LED stick and what I/O pin we'll attach it to. One of the cool things about the WS2812 is that it internally handles the Pulse-Width-Modulation (PWM) needed to control the intensity of the individual LEDs on the LED stick, so you can use any Arduino digital I/O pin to attach the data line of the LED stick. The WS2812-type addressable RGB LED sticks and strips only require 1 digital I/O pin to operate. We'll also use define statements to make setting the LED colors much easier.

```
#define NUM_LEDS 8
#define LED_PIN 5
#define RF_input_pin A0
#define gain_pin A1
#define led_Red 255,0,0
#define led_Green 0,255,0
#define led_Yellow 255,255,0
#define led_brightness 50
#define update_delay 200
```

After declaring the few variables this sketch needs, we'll then instantiate the LED stick.

```
int signal_strength = 0;
int effective_strength = 0;
int max_strength = 1000;
int LEDsToLight = 0;
int rf_gain;

Adafruit_NeoPixel pixels(NUM_LEDS, LED_PIN, NEO_GRB + NEO_KHZ800);
```

The setup() loop for the RF Probe is about as short and sweet as they come. All we need to do is start the addressable RGB LED stick, turn off all of the LED's, and then call the show function of the NeoPixel library to execute the LED clear function.

```
pixels.begin(); // INITIALIZE NeoPixel strip object

pixels.clear(); // Set all pixel colors to 'off'
pixels.show();
```

In the main loop() of the sketch, the analog voltage is read from the RF Gain potentiometer attached to the Arduino's Analog Input pin 1. Next, we'll read the RF signal strength on the Arduino's Analog Input pin 0 and we'll then use value of the Gain setting to adjust the

Arduino map() statement to provide a form of sensitivity adjustment for the RF Probe LEDs.

```
signal_strength = analogRead(RF_input_pin);  // Read Analog input
effective_strength = map(signal_strength, 0, max_strength, 0, rf_gain*4);
```

Now would be a good time to introduce the Arduino map() statement. The map() statement is a great way to modify the scale of your data. In this case, the 0 to 5 volt analog input from the AD8307 we're using to detect the RF signal is mapped from a possible digital value of 0 to 1023 into a value from 0 to 8, which is the number of LEDs on the LED stick The resulting value tells us how many LEDs on the LED stick need to be lit to indicate the RF Field Strength.

```
LEDsToLight = map(effective_strength, 0, max_strength, 0, NUM_LEDS);

  if (LEDsToLight < 0)
  {
    LEDsToLight = 0;
  }

  if (LEDsToLight > NUM_LEDS)
  {
    LEDsToLight = NUM_LEDS;
  }
```

Now that we know how many LED's to light, we can determine what color to set the lit LEDs to. Remember, we determined that the first four LEDs would always be green, the next two would always be yellow, and the last two would always be red. We start out by first clearing the previous LED setting and lighting the LEDs for the current RF signal strength reading.

```
pixels.clear();
```

Next, the colors and desired number of LEDs to light are set.

```
  pixels.setBrightness(led_brightness);

    for (int led = 0; led < LEDsToLight; led = led + 1)
    {
      if (led <= 3)
      {
    pixels.setPixelColor(led, pixels.Color(led_Green));
      }
      if (led > 3 && led <= 6)
      {
        pixels.setPixelColor(led, pixels.Color(led_Yellow));
      }
```

```
        if (led > 5)
        {
          pixels.setPixelColor(led, pixels.Color(led_Red));
        }
      }
```

Finally, the pixels.show() function is used to actually display the LED colors and the desired numbers of LEDs are then lit.

```
      pixels.show();
```

The update_delay value is used to delay between iterations of the sketch. I have it defined in the sketch at 200ms, providing 5 updates per second. You can adjust this value to suit your preferences. A value of 0 will update the LEDs on the stick as fast as the Arduino possibly can, providing a near real-time display of the signal strength. This technique could also be used to create a dynamic audio level meter for audio applications as well.

```
      delay(update_delay);
```

Enhancement Ideas

This would be an excellent project for an Altoids® mint tin and an Arduino Nano. If you used an SMA connector instead of the SO-239, you could possibly fit everything inside the Altoids tin with a small cutout for the addressable LED stick. You could also replace the LED stick with a small Organic LED (OLED) and have the RF Probe provide a digital representation of the relative field strength.

The sketch and libraries for this project are available online at www.kw5gp.com/Elektor.

Chapter 10 • Project 4 — DTMF Tone Encoder

Figure 10.1: The finished DTMF Tone Encoder.

One of the things I love to do with the Arduino is to work with components that are not well documented or understood, figure out how to use them in Arduino projects, and then de-mystify their usage for the rest of us. The Holtek 9200B DTMF (Dual Tone Multi-Frequency) Generator chip is one of those components. Hams utilizing DTMF tones to control remote devices such as repeaters, phone patches, and the like goes way back to even when I was starting out in ham radio back in the 1970's. Before the advent of cell phones, hams were the kings of the hill with their access to a phone patch on many local 2-meter repeaters that required DTMF tones to operate. Just about every handheld made today has a DTMF tone keypad, so why not create an Arduino-powered DTMF tone audio generator like the one shown in Figure 10.1? And, in the next chapter, we'll create a matching DTMF tone decoder so you can create your own ham radio remote control projects.

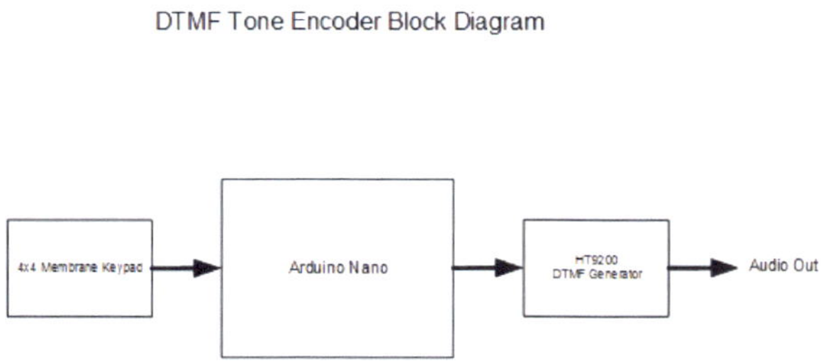

Figure 10.2: The DTMF Tone Encoder Block Diagram.

The Block Diagram for the DTMF Tone Encoder is shown in Figure 10.2. We'll use a Holtek 9200-series DTMF Generator to create the audio tones necessary for this project. The Holtek 9200 series of DTMF Generator chips is available in two basic configurations, the 9200A which is an 8-pin Serial interface version and the 9200B, a 14-pin Serial/Parallel interface version. For this project we'll use the 9200B operating in parallel interface mode since that mode is easier to work with and interface to the Arduino. The Holtek 9200 chips use a standard 3.579545 MHz crystal to create the DTMF tones. Since the Holtek 9200B chip does not have the power output capable of driving a speaker, we'll use an LM386 Audio Amplifier IC to amplify the audio output enough to drive a speaker and add a level adjustment if you'd prefer to drive the transmitter audio directly.

We'll use the standard 4 row by 4 column membrane keypad most commonly used with an Arduino. This keypad is a waterproof-type sealed plastic keypad, so it can handle fairly rugged environmental conditions without failing. This 4 x 4 keypad type has 4 row and 4 column data lines that are used to indicate which key is pressed. While the actual implementation varies between keypad models, typically the row data lines are connected to the Arduino's output pins and the column data lines are connected to the Arduino's input pins. When a key is pressed, it will connect the row pin to the column pin that the key is connected to. The Arduino will constantly scan for a key press by outputting a voltage sequentially on each of the row pins and looking for a corresponding signal on the column pin. The Adafruit_Keypad library does all this scanning work for us and tells us when and what key was pressed without you manually having to scan the keypad.

While Texas Instruments has officially announced the end of life for the LM386 audio amplifier IC, it is still readily available from the usual online sources such as eBay, Amazon, AliExpress and others. An alternative amplifier IC would be the LM4862, capable of 675mW of audio output. If you would prefer to use the LM4862 instead of the LM386, there are a number of simple amplifier designs you could use instead of the LM386 circuit used in this project. I prefer the LM386 because it can operate from 4 to 12 volts instead of the 2.7 to 5 volt limitation of the LM4862, even though the LM4862 has nearly twice the audio power output of the LM386.

Figure 10.3 shows the schematic diagram and Figure 10.4 shows the Parts List for the DTMF Tone Encoder project. The 4x4 keypad will be mounted to the top cover of a Solarbotics MegaSAFE enclosure and we'll use the DuPont-type 2.54mm header pins to connect the keypad to the circuit board. Several small holes were drilled into one of the enclosure's side panels and a small ¾-inch 8Ω speaker was glued to the inside of one of the side panels. You could also use a mini-mono phone jack if you're planning to connect the output directly to the mic input of your transmitter.

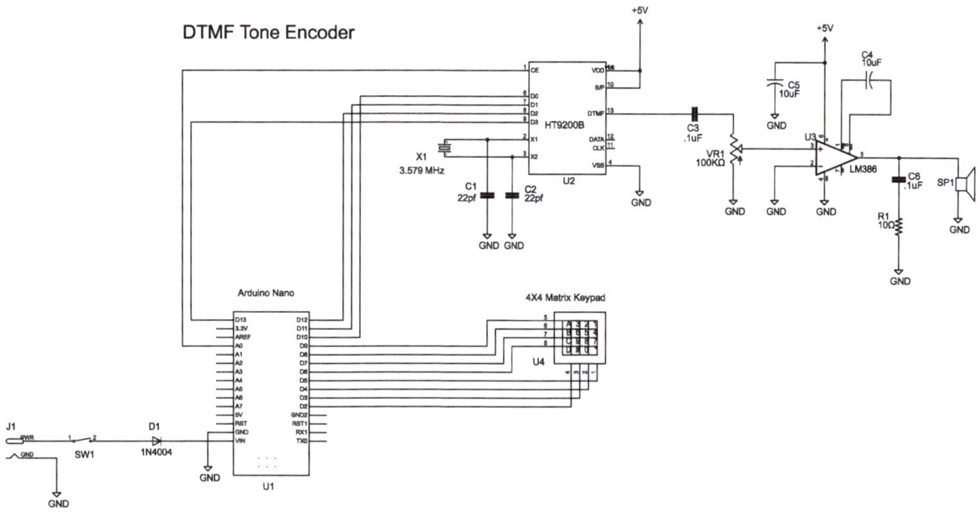

Figure 10.3 The DTMF tone Encoder schematic.

Parts List
DTMF Tone Encoder

Part	**Value**
C1, C2	22 pF ceramic capacitor
C3	0.1 µF, 25 V capacitor
C4, C5	10 µF, 25 V capacitor
C6	1 µF, 25 V capacitor
D1	1N4004 diode
J1	DC power jack
R1	10 Ω ¼ W resistor
SP1	8 Ω speaker
SW1	SPST switch
U1	Arduino Nano
U2	HT9200B Holtek touch tone encoder
U3	LM386 audio amplifier
U4	4x4 membrane keypad
VR1	100 kΩ potentiometer
X1	3.579 MHz crystal
Enclosure	Solarbotics Mega S.A.F.E.

Figure 10.4: The DTMF tone Encoder parts list.

Once the design was functional on the breadboard, it was moved to the usual 60 mm by 80 mm prototyping board (Figure 10.5) and mounted inside the Solarbotics enclosure. This is something else to think about when building your Arduino projects. I have personally chosen to standardize on the 60 mm by 80 mm prototyping boards and the Solarbotics enclosures where possible. As you design and build your Arduino projects, I recommend trying to standardize your construction methods and choice of enclosures. This ends up being a great help when you're initially working out a design idea, as you already have an idea of the construction methods and enclosure choices best suited for the project you're designing.

Figure 10.5: The DTMF Tone Encoder circuit board.

Because the Keypad Library handles all of the keypad functions for us, and the Holtek HT9200B DTMF Tone Generator chip handles all of the tone generating functions for us, the sketch for this project is literally one of those "Wait for a key press, Send the key press data to the tone generator" kind of projects. This is one of the cool things about the Arduino and it's very easy to overthink things on a project like this. This is also why I say that nearly anyone can create cool projects with the Arduino. Once you understand how to connect the various parts, the libraries will pretty much carry the day from there. This project is a perfect example of that philosophy.

Figure 10.6 shows the Flowchart we'll use to create this sketch. Even though this is a relatively straightforward sketch, with the majority of the work being done by libraries, I still find it helpful to create a flowchart to help keep my thoughts organized while writing and testing the various portions of the sketch.

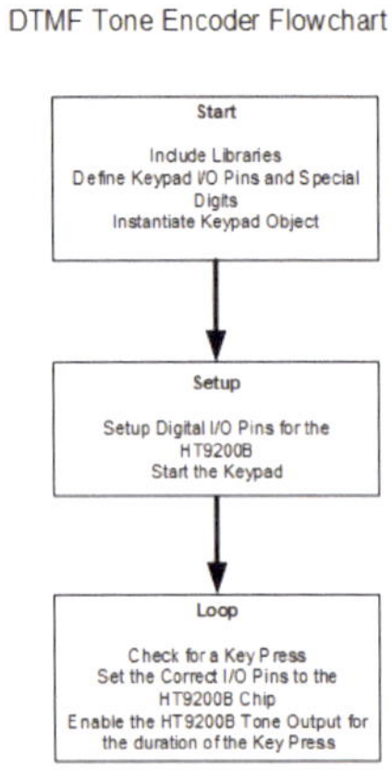

Figure 10.6: The DTMF Tone Encoder flowchart.

This sketch uses the Adafruit Keypad library with one small twist. You'll need to copy the keypad.config.h file from your Arduino>Libraries>Adafruit_Keypad>Examples>keypad_test folder to the Arduino>Libraries>AdaFruit_Keypad folder. The reason for this is that the keypad.config.h file is used as an IDE Tab in the example file, and as I mentioned earlier, I prefer to avoid using the IDE Tabs in my sketches. So, rather than having to follow the Adafruit Tab usage for the config file, if you move the file to the root of the Adafruit_Keypad library folder, it will be available to be used as a standard library "Include" statement.

You'll also need to define the keypad type that you are using. As a general rule, the KEYPAD_PID3844 definition works well with the standard 4x4 membrane-type keypads.

```
#include "Adafruit_Keypad.h"

// define your specific keypad here via PID
#define KEYPAD_PID3844  // This keypad also works with the 4x4 Membrane keypad
```

Next, let's define the Row and Column digital I/O pins that the keypad is attached to. One thing to note is that this type of keypad requires 8 digital I/O pins to support a 4x4 keypad matrix. Some of your Arduino projects may need more I/O pins than are available if you use this type of keypad in your projects. If you need more digital I/O pins than are available, you might consider using one of the other keypads available for the Arduino that use a single analog input pin such as the one shown in Figure 3.46 and 3.47 in Chapter 3 of this book, or the Adafruit Trellis-style keypads that interface to the Arduino using the I^2C bus.

```
// define the keypad pins
#define R1    5
#define R2    4
#define R3    3
#define R4    2
#define C1    9
#define C2    8
#define C3    7
#define C4    6
```

We'll add the statement that includes the keypad.config.h file we're using as a library below the above keypad pin definitions, and then also define the Zero and non-numeric keys of the key pad.

```
// leave this import after the above configuration
#include "keypad_config.h"

// Define Tone Special Digits
#define ZERO      10
#define STAR      11
#define POUND     12
#define A_KEY     13
```

```
#define B_KEY     14
#define C_KEY     15
#define D_KEY      0
```

Finally, we'll define the HT9200 digital I/O pins. Notice that we're using the Arduino's Analog A0 pin as a digital output pin since I prefer to avoid using the Arduino's digital I/O pins 0 and 1. The keypad and the other HT9200 pins use all of the remaining digital I/O pins. Fortunately, we can use the analog input pins as standard digital I/O pins as we do in this case.

```
// Define the HT9200 Pins
#define D0_PIN  10              // Data 0
#define D1_PIN  11              // Data 1
#define D2_PIN  12              // Data 2
#define D3_PIN  13              // Data 3
#define CS_PIN  14              // uses Analog pin A0 as a digital I/O pin for
Chip Select LOW = On
```

Finally, we'll instantiate the keypad object and move on into the setup() portion of the sketch.

```
Adafruit_Keypad kpd = Adafruit_Keypad( makeKeymap(keys), rowPins, colPins, ROWS,
COLS);
```

In the setup() portion of the sketch, we'll start the keypad and set the digital I/O pin modes for the HT9200 DTMF generator. Notice that we don't have to set the pin modes for the keypad. The pin modes for the keypad are set for you by the keypad library itself.

As the last step of the setup(), we'll make sure the DTMF audio tones are turned off by calling the turn_off_tone() function.

```
kpd.begin();

  // Set the pin modes for the HT9200B
  pinMode(D0_PIN, OUTPUT);
  pinMode(D1_PIN, OUTPUT);
  pinMode(D2_PIN, OUTPUT);
  pinMode(D3_PIN, OUTPUT);
  pinMode(CS_PIN, OUTPUT);
  turn_off_tone();
```

In the main loop, we start out by scanning the keypad to look for a key press by calling the library tick() function. This function will scan the keypad and return the digit that was pressed, if any.

```
kpd.tick();

  while (kpd.available()) // Check if keypad button is pressed
  {
    keypadEvent e = kpd.read(); // get the keypad button
    key_char = (char)e.bit.KEY;
```

If there is a key pressed, we call the send_key_tone() function which will enable the Chip Select (CS) pin on the HT9200B chip. The tone will continue to be sent as long as the key is pressed.

```
    if (e.bit.EVENT == KEY_JUST_PRESSED)
    {
      send_key_tone();
    }
```

When the key is released, we call the turn_off_tone() function that will disable the CS pin on the HT9200B, thereby turning off the tone.

```
else if (e.bit.EVENT == KEY_JUST_RELEASED)
    {
      turn_off_tone();
    }
```

There are three main functions used by this sketch, the send_key_tone() function, the convert_key() function, and the turn_off_tone() function. The send_key_tone() function will call the convert_tone_function() to convert the ASCII key value sent by the keypad library to the 4 pin digital I/O representation that is needed by the HT9200B to send the correct tones.

```
void convert_key()
{

  digit = int(key_char);

  if (digit >= 49 && digit <= 57)
  {
    digit = digit - 48;
  }
  else
  {
    switch (key_char)
    {
      case ‚0‘:
        digit = ZERO;
```

```
        break;

      case ,A‘:
        digit = A_KEY;
        break;

      case ,B‘:
        digit = B_KEY;
        break;

      case ,C‘:
        digit = C_KEY;
        break;

      case ,D‘:
        digit = D_KEY;
        break;

      case ,*‘:
        digit = STAR;
        break;

      case ,#‘:
        digit = POUND;
        break;
    }
  }
}
```

Once the key press is converted to the 4 pin representation, it is sent to the HT9200B and the CS pin is enabled, starting the tone transmission.

```
void send_key_tone()
{
  convert_key();

  // Turn on the tone for the selected key
  digitalWrite(D0_PIN, bitRead(digit, 0));
  digitalWrite(D1_PIN, bitRead(digit, 1));
  digitalWrite(D2_PIN, bitRead(digit, 2));
  digitalWrite(D3_PIN, bitRead(digit, 3));
  delay(100);
  digitalWrite(CS_PIN, LOW);           // Enable the tone encoder
}
```

When the key is released, we call the turn_off_tone() function to disable the CS pin and set the HT9200 data in pins back to LOW and wait for the next key press.

```
void turn_off_tone()
{
  // Turn off the tone
  digitalWrite(CS_PIN, HIGH);          // Disable chip (req if parallel)
  digitalWrite(D0_PIN, LOW);
  digitalWrite(D1_PIN, LOW);
  digitalWrite(D2_PIN, LOW);
  digitalWrite(D3_PIN, LOW);
  delay(100);
}
```

Enhancement Ideas

As with many of my book projects, I like to leave a little room for you to add your own personal touches to the projects, add features, etc. There are several things I would do to enhance this particular project to really give it that extra little kick. You could add a color TFT display to display the actual keys pressed, as well as using one of the Arduino's I/O pins such as Analog pin A1 used as a digital I/O pin to trigger a relay that can be used to drive the transmitter's push to talk (PTT).

The sketch and libraries for this project are available online at www.kw5gp.com/Elektor.

Chapter 11 • Project 5 — DTMF Tone Decoder

Figure 11.1: The finished DTMF Tone Decoder.

Figure 11.2: The DTMF Tone Decoder TFT display.

If you're going to have a DTMF tone encoder project, naturally you'll need a DTMF tone decoder project (Figures 11.1 and 11.2) to go along with it. Since many of today's handheld and mobile transceivers have the capability to send DTMF tones, it would be great if you could use those tones to control things remotely. Once again, there's a chip for that, but outside of the datasheet, it too is relatively undocumented when it comes to use in Arduino projects.

The Zarlink (now Microsemi Corporation) MT8870DE is an Integrated DTMF Receiver in a single 18-pin DIP chip. This chip incorporates both a bandsplit filter to handle the high and low frequency tone group filtering and can detect all 16 DTMF tone pairs and output a corresponding 4-bit code. Like the Holtek 9200 chip used in the DTMF Tone Encoder project, the MT8870 chip uses a 3.57945-MHz crystal to drive the chip's internal oscillator and frequency counter.

The block diagram for the DTMF Tone Decoder project is shown in Figure 11.3. As mentioned above, we'll use the Zarlink MT8870 chip to decode the DTMF tones. The MT8870 outputs a 4-bit digital representation of the received tone on its output data lines and will set the StD (Delayed Steering Output) pin HIGH when it receives a valid tone pair. We'll connect those MT8870 pins to digital input pins on the Arduino. We'll wait for a valid tone pair as indicated by the StD pin and decode the received digit. If the received digit is the digit selected to turn the relay on or off, the relay is set to the on or off condition specified by the received digit. Once the relay is turned on, the sketch will leave the relay on and monitor the received tones until it detects the specified digit to turn off the relay. This project only uses a single tone pair to turn the relay on and off. For this project. an asterisk (*) turns the relay on and receiving a number (also known as a pound) sign (#) will turn the relay off. You can also modify the sketch to require multiple tone sequences to turn the relay on and off as well. We'll display the decoded received digits on a color TFT display, along with the relay status information.

Notice that the only Arduino library we'll be using is the Adafruit TFT display library. While it turns out that here is an Arduino library for the MT8870, because interfacing the Arduino to the MT8870 chip appeared to be so easy, I didn't think to look for one when I was writing the sketch for this project. So, we can add one more thing to our personal list of Arduino design rules to make things even easier... always look for a library that can do what you need and possibly save yourself from having to do it yourself. Fortunately, in this case my oversight didn't overcomplicate things or affect the time it took to create the sketch, but the library could very well have had extra features I might have wanted to include in the project. Fortunately for me, that wasn't the case this time.

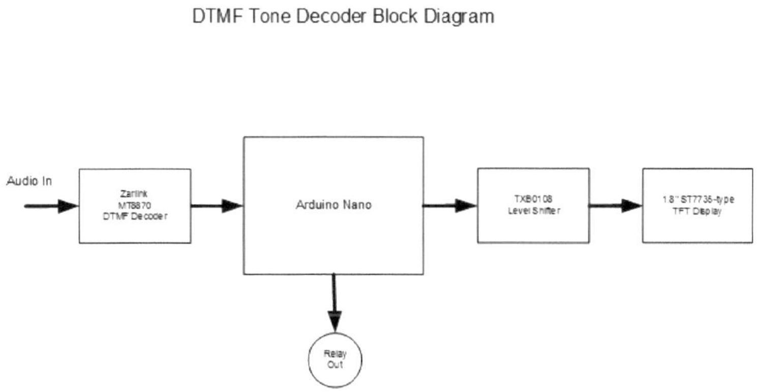

Figure 11.3: The DTMF Tone Decoder block diagram.

Figure 11.4 shows the schematic diagram and 11.5 shows the parts list for the DTMF Tone Decoder project. As with most of my projects involving a TFT display, the ST7735-type TFT display is mounted to the inside of the top cover of a Solarbotics MegaSAFE enclosure using 2 mm hardware. The rest of the components are mounted on a 60 mm by 80 mm copper-clad prototyping board as shown in Figure 11.6. The completed circuit board is then mounted inside the MegaSAFE enclosure. Mini-mono audio jacks are mounted to the side panel of the enclosure for the audio input and the relay contact outputs.

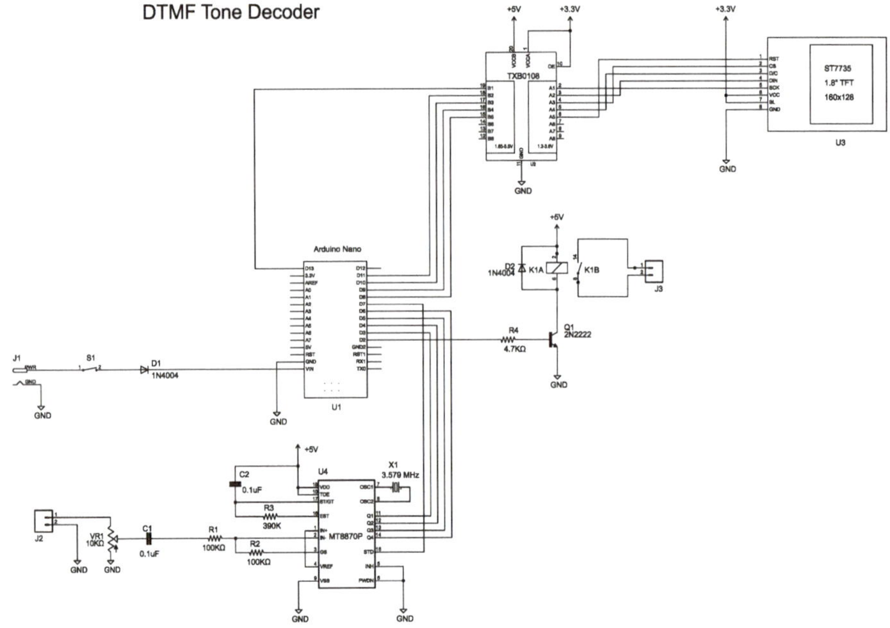

Figure 11.4: The DTMF Tone Decoder schematic diagram.

Parts List
DTMF Tone Decoder

Part	Value
C1, C2	0.1 µF ceramic capacitor
D1, D2	1N4004 diode
J1	DC power jack
J2, J3	mini mono phone jack
K1	5 V reed relay
Q1	2N2222A transistor
R1, R2	100 kΩ 1/8W resistor
R3	390 kΩ 1/8W resistor
R4	4.7 kΩ 1/8W resistor
S1	SPST switch
U1	Arduino Nano
U2	8-channel level shifter module (TXB0108 type)
U3	ST7735-type 1.8" color TFT Display
U4	MT8870 Microchip/Zarlink tone decoder IC
VR1	10 kΩ potentiometer
X1	3.579 MHz crystal
Enclosure	Solarbotics Mega S.A.F.E.

Figure 11.5: The DTMF Tone Decoder parts list.

Figure 11.6: The DTMF Tone Decoder circuit board.

The Flowchart for the DTMF Tone Decoder project is shown in Figure 11.7. This is another one of those cases where the Arduino, TFT Display, MT8870 chip, and the Arduino libraries do 90% of the work for us and is yet another example of why the Arduino is so popular among new builders. The hardest part of building projects like this is figuring out how to wire everything together. The sketches for projects such as this one are literally not much more than "program glue" between the various modules and libraries. As you'll see in the sketch, we do nearly exactly what the flowchart shows, and in the exact same sequence.

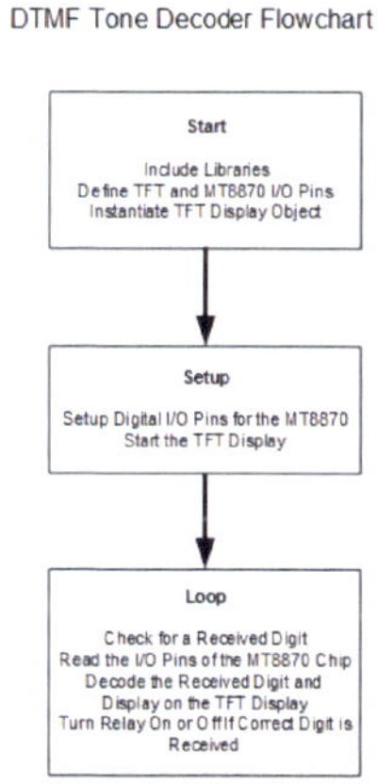

Figure 11.7: The DTMF Tone Decoder flowchart.

We start out the sketch as usual, including the Adafruit libraries for the ST7735-type color TFT display module, then defining the digital I/O pin numbers used by the TFT display, MT8870 chip, and the relay. Finally, we'll define the font sizes and the line length of the TFT display used to display the received DTMF digits. We use this line length value to provide a continuously scrolling display showing the last 8 DTMF digits we've received.

```
#define TFT_CS     10  // Assign the TFT CS to pin 10
#define TFT_RST    8   // Assign the TFT RST to pin 7
#define TFT_DC     9   // Assign the TFT DC to pin 6
```

```
#define StD_pin 7
#define Q1_pin 3
#define Q2_pin 4
#define Q3_pin 5
#define Q4_pin 6
#define relay_pin 2

#define text_font_size 3    // The TFT text display font size
#define relay_font_size 2
#define status_font_size 1  // The font size for the status line
#define tft_line_length 8
```

Next, we'll declare the variables used by the sketch and clear the rotating buffer variable we'll use to display the received digits. Then we'll instantiate the TFT display object before moving on to the setup() section of the sketch.

```
bool previous_state = LOW;
bool Tone_Decoded = LOW;
bool relay_active = LOW;

int Q1;
int Q2;
int Q3;
int Q4;
int digit_value;
char digit;

String tft_data = «»;  // String variable to hold tft text data

// Instantiate the TFT Display
Adafruit_ST7735 tft = Adafruit_ST7735(TFT_CS,  TFT_DC, TFT_RST);  // Initialize
the TFT display instance
```

In the setup() section, we'll set the digital I/O pin modes for the MT8870 data and StD pins as well as the relay pin. We turn off the relay and then display a brief startup "Splash Screen" on the TFT display. After a 5 second delay, the TFT display screen is cleared and we move on to the main sketch loop().

```
pinMode(StD_pin, INPUT);
  pinMode(Q1_pin, INPUT);
  pinMode(Q2_pin, INPUT);
  pinMode(Q3_pin, INPUT);
  pinMode(Q4_pin, INPUT);
```

```
  pinMode(relay_pin, OUTPUT);

  digitalWrite(relay_pin, LOW);

  // Display the splash screen
  tft.initR(INITR_18BLACKTAB);    // initialize a 1.8" TFT with ST7735S chip,
black tab
  delay(tft_delay);
  tft.fillScreen(ST7735_BLUE);
  delay(tft_delay);
  tft.setRotation(1);             // Set the screen rotation
  delay(tft_delay);
  tft.setTextWrap(false);         // Turn off Text Wrap
  delay(tft_delay);
  tft.setTextSize(text_font_size);// Set the Font Size
  delay(tft_delay);
  tft.setTextColor(ST7735_WHITE); //Set the Text Color
  delay(tft_delay);
  tft.setCursor(40, 10);          //Set the Cursor and display the startup screen
  delay(tft_delay);
  tft.print("KW5GP");
  delay(tft_delay);
  tft.setTextSize(2);
  delay(tft_delay);
  tft.setCursor(45, 60);
  delay(tft_delay);
  tft.print("Arduino");
  delay(tft_delay);
  tft.setCursor(30, 80);
  delay(tft_delay);
  tft.print("TouchTone");
  delay(tft_delay);
  tft.setCursor(45, 100);
  delay(tft_delay);
  tft.print("Decoder");
  delay(tft_delay);

  delay(5000);
  clear_display();
```

In the sketch loop(), we continuously loop, waiting for the MT8870's StD pin to go HIGH, indicating that we have received a valid tone pair. Note that since StD remains HIGH as long as the tone pair is received, we have the sketch ignore the StD pin until it has gone LOW, indicating that it is no longer is receiving the tone pair. We do this so that we don't receive and display the same digit multiple times. Once the tone is no longer received, we reset the flag and wait for the next valid tone pair.

```
void loop()  // Main Program Loop
{
  // Read the StD pin

  Tone_Decoded = digitalRead(StD_pin);

  if (Tone_Decoded == HIGH)
  {
    if (previous_state == LOW)
    {
      decode_tone();
      update_text();
      previous_state = HIGH;
    }
  }
  else
  {
    if (previous_state == HIGH)
    {
      previous_state = LOW;
    }
  }
}
```

As you can see, the again the sketch loop() is rather small for this project. Once a valid tone pair is indicated, we call the decode_tone() function to decode the digit received and then call the update_text() function to display the received digit on the TFT display. The decode_tone() function will also turn the relay on and off if it receives the correct digit.

There are six functions used in this sketch. Breaking your sketch out into functions allows you to have a much smaller, and therefore a more easily read and understood sketch loop(). This also allows you to reuse these functions in other projects, saving you from having to recreate these functions when you use them in those other projects.

The first function is the clear_display() function. This is actually a bit of a misnomer, as I like to use this function to not only clear the TFT display, but to also display a template. You may have noticed by now that the Arduino UNO and Nano aren't fast enough to clear the TFT display instantaneously. By using a function similar to the clear_display() function and displaying a template that doesn't change, I only have to update the text that changes on the display, allowing for faster display of the data without having to clear the entire display between updates.

```
void clear_display()  // Clears the TFT display
{
  tft.fillScreen(ST7735_BLUE);
  delay(tft_delay);
```

```
  tft.setTextSize(status_font_size); // Set the Font Size
  delay(tft_delay);
  tft.setCursor(40, 25);
  delay(tft_delay);
  tft.print("Decoded Digit");
  delay(tft_delay);
  update_relay_status();
  delay(tft_delay);
  tft.setTextSize(status_font_size);
  delay(tft_delay);
  tft.setCursor(40, 115);
  tft.print("Decoder Ready");
  delay(tft_delay);
  tft.setTextSize(text_font_size); // Set the Font Size
  delay(tft_delay);
  tft.setCursor(10, 40);
  delay(tft_delay);
}
```

The decode_tone() function will read the 4 data lines coming from the MT8870 and return the digit that was received. It will also turn the relay on and off if the proper digit is received. We'll use a switch...case() statement to display the correct digit and to control the relay operation.

```
void decode_tone()
{
  // Decodes the 4 bits of the Tone Decoder and returns the digit as a character

  // Read the decoded tone bits
  Q1 = digitalRead(Q1_pin);
  Q2 = digitalRead(Q2_pin);
  Q3 = digitalRead(Q3_pin);
  Q4 = digitalRead(Q4_pin);

  // Now convert it to a digit

  digit_value = (Q4 * 8) + (Q3 * 4) + (Q2 * 2) + Q1;

  /*    Decoded Values
        1-9 = Actual Value
        A = 13
        B = 14
        C = 15
        D = 0
        Asterisk (*) = 11
        # = 12
```

```
        0 = 10
  */

  switch (digit_value)
  {
    case (13):
      digit = 'A';
      break;

    case (14):
      digit = 'B';
      break;

    case (15):
      digit = 'C';
      break;

    case (0):
      digit = 'D';
      break;

    case (11):
      digit = '*';
      set_relay();
      break;

    case (12):
      digit = '#';
      reset_relay();
      break;

    case (10):
      digit = '0';
      break;

    default:
      digit = digit_value + ,0';
      break;

  }
}
```

The update_text() function will update the text area on the TFT display that is used to show the received digits. This function clears only a small portion of the TFT screen and allows for faster and smoother updating of the received digits.

```
void update_text() // Updates the tft text on the display
{
  if (tft_data.length() >= tft_line_length)  // Scroll one line of characters
  {
    tft_data = tft_data.substring(1);  // Remove the first character
    tft.setCursor(10, 40);  // Move the cursor to the top line
    tft.fillRect(0, 40, 159, 30, ST7735_BLUE); // Clear the tft data area
    tft.print(tft_data);    // Print the tft data string variable
  }
  tft_data = tft_data + digit;  // Add the character to the tft data string
  tft.print(digit);             // display the received character
}
```

The set_relay() function is used to turn on the relay and calls the update_relay_status() function to update the relay status shown on the TFT display.

```
void set_relay()
{
  digitalWrite(relay_pin, HIGH);
  relay_active = true;
  update_relay_status();
}
```

The reset_relay() function is used to turn off the relay and calls the update_relay_status() function to update the relay status shown on the TFT display.

```
void reset_relay()
{
  digitalWrite(relay_pin, LOW);
  relay_active = false;
  update_relay_status();
}
```

The update_relay_status() function is used to update the relay status that is displayed on the TFT display. This function only clears the area where the relay status information is displayed, allowing for smoother and faster updating of the TFT display.

```
void update_relay_status()
{
  tft.fillRect(0, 40, 159, 60, ST7735_BLUE); // Clear the tft data area
  delay(tft_delay);
  tft.setCursor(30, 80);
  delay(tft_delay);
  tft.setTextSize(relay_font_size);
  delay(tft_delay);
  tft.print("Relay ");
```

```
    delay(tft_delay);
    if (relay_active)
    {
      tft.print("On");
    }
    else
    {
      tft.print("Off");
    }
    delay(tft_delay);
    tft.setTextSize(text_font_size);
    delay(tft_delay);
    tft.setCursor(10, 40);
    delay(tft_delay);
    tft.print(tft_data);
    delay(tft_delay);
  }
```

Troubleshooting

Rather than transmit tones over the air to a receiver to test the finished project, you can use an Internet-based Tone Generator similar to the one at https://onlinetonegenerator.com/dtmf.html as shown in Figure 11.8. Using a USB audio dongle attached to my workstation, I fed the audio out from the workstation to the input of the DTMF Tone Decoder project. I used my lab oscilloscope to set the level to approximately 250 mV peak-to-peak. Note that you can't use a multimeter to set the signal level correctly since a multimeter is made to read 60 Hz or 50 Hz on the AC setting and the DTMF tones are in the range of 697 Hz to 1633 Hz. In lieu of using an oscilloscope, you can start at the minimum level on the input level potentiometer VR1 and slowly increase the level until the tone pairs are reliably detected.

Figure 11.8: Screenshot of the Online DTMF Tone Generator in action.

Enhancement Ideas

This is one of those projects that I think can be endlessly adapted to add new functions and relay controls. The first thing I would do is to modify the sketch so that it supports a two or three digit sequence to turn the relay on and off. Another thing would be to add multiple relays and multiple sequences to control several relays rather than just the one. Since you can use the Arduino Nano's analog input pins as digital I/O pins, you could add as many as 8 additional relay outputs by using the Nano's analog pins A0 thru A7. Finally, I would go back and modify the sketch to use the Arduino MT8870 library, just to keep everything consistent with my personal Arduino design philosophy.

The sketch and libraries for this project are available online at www.kw5gp.com/Elektor.

Chapter 12 • Project 6 — Waveform Generator

Figure 12.1: The finished Waveform Generator project.

A Waveform Generator similar to the one shown in Figure 12.1 is another piece of test equipment that's handy to have around. If you're going to troubleshoot or work with audio circuits, it's always nice to have a stable and clean audio source available. Since I can't whistle, I'm pretty much out of luck when it comes to testing audio circuits without something to generate the audio for me. While I could use the tone() function on the Arduino to drive a digital I/O pin, this solution wouldn't be the best choice for several reasons. First, the tone() method generates a square wave, which isn't very good for testing audio signals due to the sharp rise and fall times of the waveform, as well as the harmonics caused by these sharp transitions. Then there is the question of accuracy, since the frequency is determined by a timer register inside the Arduino chip and driven by software. Yes, you could use filtering to smooth the square waves into sine waves, but again, we're starting to make this design overly complex. There has to be a better way to do this process in a much simpler way.

Fortunately, there is a much simpler solution available. I was in total amazement when I discovered the Direct Digital Synthesizer (DDS) modules. Prior to this, the only way I knew how to generate a stable frequency was to use a crystal oscillator. The downside to crystal oscillators is that they are only capable of delivering a single frequency unless you manually divided the frequency and now the chip count keeps going higher and higher for a simple project. The next step beyond that was some form of variable oscillator using something along the lines of the venerable NE555 Timer chip as an oscillator, or a discrete component inductive-capacitive oscillator. Unfortunately, neither solution is highly accurate and subject to frequency drift.

The DDS changes all that. Now you can generate highly accurate and stable frequencies over a wide range, all with a single chip. For example, the Analog Devices AD9833 Programmable Waveform Generator (Figure 12.2) used in this project has a frequency output of 0 to 12.5 MHz in 0.1-Hz steps through the use of a 28-bit frequency control register. Other DDS modules can generate frequencies up to 400 MHz, and my preferred DDS for RF frequencies, the AD9850, has a maximum frequency of 62.5 MHz with a resolution of 0.0291 Hz and a 32-bit frequency tuning register.

To show just how much of a game-changer the DDS module is, we'll walk through the concept, design, and breadboard prototyping phases of this project in detail.

Figure 12.2: The AD9833 Programmable Waveform Generator module.

This is one of those projects that kept evolving as I worked with it on the breadboard. Had I just soldered up the original design, I would have spent more time soldering (and desoldering) than I would have actually spent building the project. This is a perfect example of why building and testing your initial concepts and designs on a breadboard is the preferred way to go.

My initial design criteria called for a waveform generator that could generate a sine, square, and triangle wave from 1 Hz to 20 KHz or better. Since the Arduino can only generate square waves on the digital I/O pins, and has no Digital to Analog converter onboard, I would have to come up with a way to generate the waveforms externally. That sounded simple enough in theory, I had already found numerous circuits that could do this in the Arduino Playground on the Arduino.cc website.

The simplest design (Figure 12.3) called for a series of resistors connected to the Arduino's digital I/O pins that would create what is known as a "resistor-to-resistor ladder network" (R2R) digital to analog (D/A) converter. This design uses a series of resistors connected across 8 digital I/O pins on the Arduino to generate the analog output. In effect, this creates an 8-bit digital to analog converter, and the Arduino would be used to generate the actual waveforms by using 8 digital I/O pins as the input to the R2R D/A network.

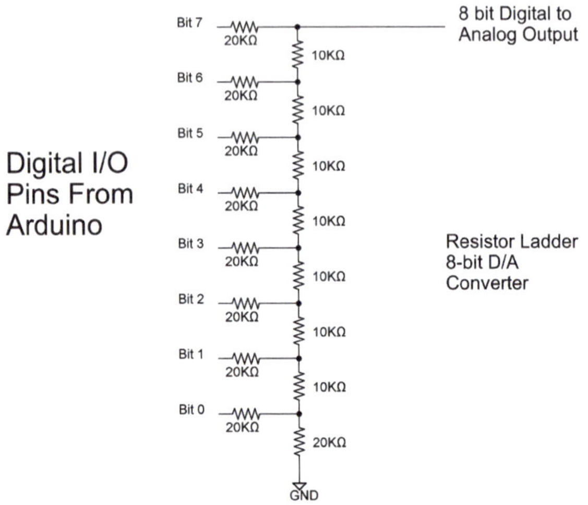

Figure 12.3: A resistive-ladder Digital to Analog Converter.

The test sketch for this was simple, and there were plenty of example sketches using a table stored in flash memory to generate the sine wave. To generate a triangle wave, all I had to do was increment and decrement the data sent to the digital I/O pins. Sounds easy enough, right?

Well, at 200 Hz (Figure 12.4), everything was perfect. However, as the frequency went up, the waveform started to distort. At 6 KHz (Figure 12.5), there was a noticeable stair-stepping to the waveform, and at 13 KHz (Figure 12.6), the sine wave was seriously distorted. At 27 kHz (Figure 12.7), the sine wave was so distorted that it was more like a bad square wave than a sine wave. There was no reason to try higher frequencies as it was obvious this solution wouldn't work out. It was the same way with a triangle wave, except the distortion was very noticeable at 8 kHz, and at 12.5 kHz the triangle wave was distorted to where it began to look more like a very bad sine wave than a triangle wave.

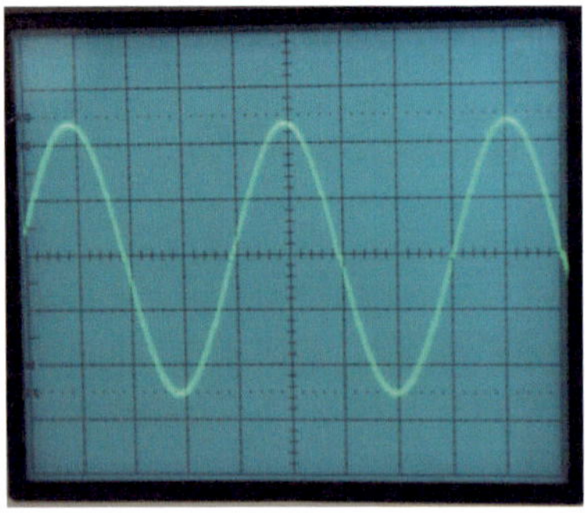

Figure 12.4: Resistive ladder D/A sinewave output at 200 Hz.

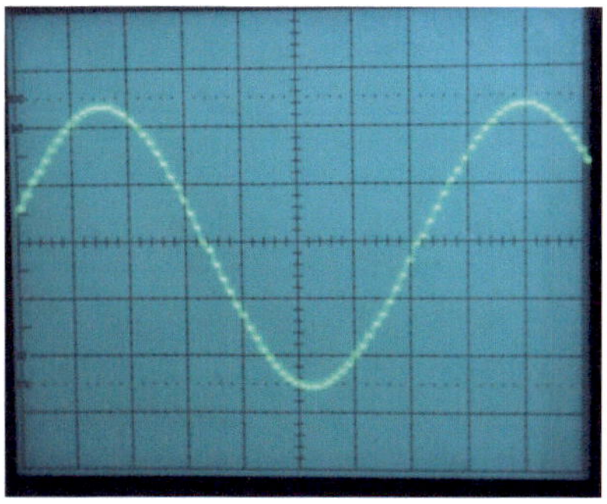

Figure 12.5: Resistive ladder D/A sinewave output at 6 KHz.

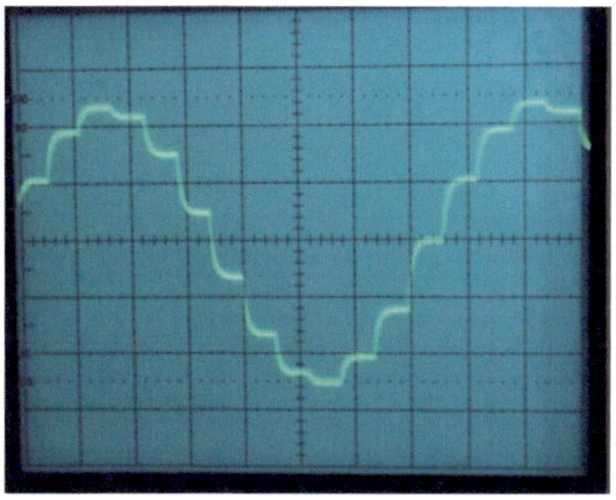

Figure 12.6: Resistive ladder D/A sinewave output at 13 KHz.

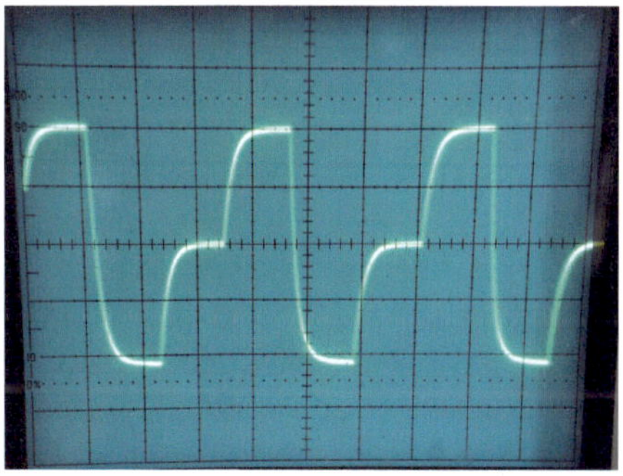

Figure 12.7: Resistive ladder D/A sinewave output at 27 kHz.

I tried multiple versions of the sketch and did all that I could think of to optimize the waveforms, including direct calculation of the sine wave rather than using a pre-defined table. The results were every bit (no pun intended) as bad, if not worse. Even using the Arduino's digital I/O Port Register I/O method that outputs to 8 digital I/O pins with a single digitalWrite() statement didn't improve things very much. It was pretty obvious that for any frequency above 800 Hz, this design wasn't going to provide a clean, stable waveform using the Arduino UNO or Nano. I felt that this design just wasn't working out for what should be a simple waveform generator project, so it was time to fall back and regroup.

Figure 12.8: The MCP4725 12-bit Digital to Analog Converter module.

The next design called for replacing the resistor ladder D/A with an MCP4725 I²C 12-bit D/A module. My thinking was that the MCP4725 D/A (Figure 12.8) would provide 4 additional bits of resolution and would offload some work from the Arduino since the MCP4725 uses the I²C bus for communication. Some quick rewiring on the breadboard and the new design was ready to test. As it turned out, the waveforms were slightly less distorted, but the Arduino Nano and MCP4725 couldn't generate the waveforms fast enough to produce anything usable above 800 Hz. It was fairly obvious that this design wasn't going to cut it either.

Things were not looking good for the Waveform Generator project at this point. I had exhausted all of the traditional ways to generate waveforms with the Arduino, and there was only one design idea left to try.

I had recently begun to experiment with the Direct Digital Frequency Synthesis (DDS) modules as part of another crazy project that was successful in getting the Ten-Tec Rebel Model 506 CW-only Transceiver to run the JT65 digital mode, as well as working on some ideas for a DDS-based Antenna Analyzer. Because of those project ideas, I had a few different DDS modules lying around the lab. I had been avoiding them after skimming the datasheet because it seemed they were rather complex, and I just hadn't taken the time to dig into them yet. Now was as good a time as any, so I dug back into the datasheets to see if I could use a DDS module for this project. I settled on trying to use the Analog Devices AD9833 Programmable Waveform Generator module. The reason that I chose the AD9833 was that of all the DDS modules available at a reasonable price, the AD9833 was one of the few that would directly generate triangle waves in addition to the standard sine and square wave outputs of the other DDS modules.

Direct Digital Frequency Synthesis is a relatively new technology that is rapidly finding its way into circuit designs that call for generating high-quality varying frequency signals. Using a reference clock, a DDS can be used to generate very precise and stable signals that can be used as a VFO for a transmitter or as a receiver IF mixer signal input, for example. What makes the DDS modules so appealing is that they use a standard SPI bus interface for communication with the Arduino or other microcontrollers. Controlling the Analog Devices DDS modules uses the same basic internal register format between the various DDS modules, so once you get one DDS design working, you can switch between the various DDS modules rather easily. The major difference between the Analog Devices DDS modules is their upper frequency limit, number of bits in the frequency and phase control registers, and the types of waveforms they can generate.

The Ten-Tec Rebel Model 506 QRP Transmitter used an AD9834 to generate the transmitted CW signal and the receiver IF mixing signal. The AD9834 can generate clean and stable waveforms up to 37.5 MHz, perfect for use in Amateur HF transmitter or receiver designs. The AD9850 DDS can generate signals up to 62.5 MHz and the AD9851 can generate signals up to 90 MHz. However, since the AD9834, AD9850, and AD9851 don't generate triangle waves, and this project's initial design called for generating sine, square, and triangle waves, with no requirement for generating waveforms higher than 20 KHz, the AD9833 seemed to be the best candidate for the job.

The Analog Devices AD9833 is a digitally programmable waveform generator, capable of generating sine, square, and triangle waveforms from 0 to 12.5 MHz. It has dual 28-bit frequency registers and can achieve a resolution of 0.1 Hz with a 25-MHz reference clock rate. Even higher resolutions can be achieved by using a lower frequency for the reference clock. The AD9833 also has two 12-bit phase registers, allowing the phase of the output signal to be shifted from 0 to 720° (0 to 2п). The AD9833 uses a 10-bit D/A converter to generate the output waveform. Sine waves are generated using an on-chip sine-wave lookup table in internal Read-Only Memory (ROM), requiring no external calculations to generate smooth, 10-bit sine waves across the entire frequency range. The output signal from the AD9833 is approximately 0.6 V peak-to-peak.

Without an Arduino library, you'd have to go through some hoops to communicate with a DDS module. In the case of the AD9833, all of its functions are controlled through a single 16-bit multi-function control resister. The good news is that with the AD9833 library, we don't have to figure out how to deal with the control register. In fact, we don't have to do much other than start the DDS, set the desired waveform type, and set the frequency. That's really all there is to it.

Finally, we have a basic circuit that looks like it can do what our design calls for. Creating the Block Diagram (Figure 12.9), we'll use an AD9833 to generate the waveforms, an SPDT center-off switch to select between sine, square, and triangle waves, and a rotary encoder to control the output frequency and the frequency step rate. We'll use an LM386 Audio Amplifier chip to amplify the output of the DDS module to provide an output signal strong enough to drive a speaker. We'll also have a potentiometer on the LM386 to control the output level.

We'll add an ST7735-type color TFT display to show the frequency, waveform type, and rotary encoder frequency step size as shown in Figure 12.10.

```
#define Square_wave 6          // Define Square Wave Switch Input
#define Triangle_wave 7        // Define Triangle Wave Switch
#define min_freq 50            // Define the minimum frequency as 50Hz
#define max_freq 100000        // Define the maximum frequency as 1 MHz
```

Then we'll assign the digital I/O pins for the rotary encoder and its pushbutton switch, along with a 500-ms debounce delay for the pushbutton switch. Then, we'll set up the starting frequency for the AD9833 as well as set up an array variable to hold the various frequency step sizes that will be controlled by the rotary encoder's pushbutton switch. We'll also set the waveform type to a sine wave and the remaining variables needed by the sketch.

```
const int encoder_PinA = 2;           // Pin A of the Rotary Encoder
const int encoder_PinB = 3;           // Pin B of the Rotary Encoder
const int encoder_Switch_Pin = 4;     // Encoder Pushbutton Switch input pin
boolean encoder_Switch;
const int debounceInterval = 500;     //disable step change for 500 ms between
encoder pushbutton press

double freq = 1000;                   //32-bit global frequency variable
double step_size[] = {1, 10, 100, 1000, 10000}; // The step rate selected by the
rotary encoder switch
double previous_frequency = freq;
int wave_type, previous_type = 0;
int encoder_Pos = 0;
int old_encoder_Pos = 0;
double current_step_size = step_size[2];
int step_index = 2;
unsigned long int delay_time = 0;
String wave_mode = «   Sine»;
String spaces = «»;
```

Finally, we'll instantiate the DDS, TFT display, and rotary encoder objects before continuing on to the setup() section of the sketch.

```
// Instantiate the AD9833
MD_AD9833  AD(FSYNC);

// Instantiate the TFT Display
Adafruit_ST7735 tft = Adafruit_ST7735(TFT_CS,  TFT_DC, TFT_RST);  // Initialize
the TFT display instance

// Instantiate the Rotary Encoder
Encoder Enc(encoder_PinA, encoder_PinB);
```

In the setup() loop, we'll set the pin modes for the rotary encoder pushbutton switch and the SPDT Mode switch before initializing the TFT display and the AD9833 before displaying a brief startup message for 5 seconds, then clearing the display.

```
  pinMode (encoder_Switch_Pin, INPUT_PULLUP);
  pinMode(Square_wave, INPUT_PULLUP);      // Set the Square_wave Control Pin as
an Input
  pinMode(Triangle_wave, INPUT_PULLUP);    // Set the Triangle_wave Control Pin as
an Input

  // Display the splash screen
  tft.initR(INITR_18BLACKTAB);           // initialize a 1.8" TFT with ST7735S chip,
black tab
  delay(tft_delay);
  tft.fillScreen(ST7735_BLUE);
  delay(tft_delay);
  tft.setRotation(1);                    // Set the screen rotation
  delay(tft_delay);
  tft.setTextWrap(false);                // Turn off Text Wrap
  delay(tft_delay);
  tft.setTextSize(text_font_size);       // Set the Font Size
  delay(tft_delay);
  tft.setTextColor(ST7735_WHITE);        // Set the Text Color
  delay(tft_delay);
  tft.setCursor(40, 10);                 // Set the Cursor and display the startup
                                           screen
  delay(tft_delay);
  tft.print("KW5GP");
  delay(tft_delay);
  tft.setTextSize(2);
  delay(tft_delay);
  tft.setCursor(40, 60);
  delay(tft_delay);
  tft.print("Arduino");
  delay(tft_delay);
  tft.setCursor(35, 80);
  delay(tft_delay);
  tft.print("Waveform");
  delay(tft_delay);
  tft.setCursor(30, 100);
  delay(tft_delay);
  tft.print("Generator");
  delay(tft_delay);

  AD.begin(); // Start the DDS
  AD.setFrequency(MD_AD9833::CHAN_0, freq);
```

```
  delay(5000);
  clear_display();
```

You'll notice yet again that the sketch loop() is very short, consisting of only three function calls. This allows us to move all of the heavy lifting into the functions and compartmentalizing the sketch operation to help make coding and troubleshooting easier.

In the sketch loop(), we'll call the read_encoder() function to see if the rotary encoder position has been changed. If it has, we'll increase or decrease the frequency variable accordingly and call the AD9833 library function to set the new frequency. We'll go through all of the functions used in this sketch in a bit, but for now we'll continue on with the main sketch loop(). Next, we'll check to see if the rotary encoder pushbutton switch has been pressed, and if it has, we'll update the value of the frequency step variable with the next value in the array, allowing us to quickly set the desired frequency. Each press of the rotary encoder's pushbutton switch will change to the next step value in the array and the TFT display will be updated to reflect this change.

```
void loop()
{
  read_encoder();      // Read the encoder and set the Tx/Rx frequencies

  read_pushbutton();   // Read the Encoder Pushbutton switch and set Fast/Slow
tuning mode

  read_wave_type();    // Read the Wave Type Switch and set the Wave Mode

}  // End Main Loop
```

The clear_display() function will clear the TFT display and display the initial screen showing the frequency and frequency step size. This function is also used to create the display template, so that future display updates only need to erase a portion of the display area.

```
void clear_display()  // Clears the TFT display
{
  tft.fillScreen(ST7735_BLUE);
  delay(tft_delay);
  tft.setTextSize(status_font_size);
  delay(tft_delay);
  tft.setCursor(60, 25);
  delay(tft_delay);
  tft.print("Frequency");
  delay(tft_delay);
  tft.setTextSize(freq_font_size);
  delay(tft_delay);
  tft.setCursor(5, 50);
```

```
delay(tft_delay);
// Add spaces to center the frequency display
spaces = ("   ");
if (freq < 100)
{
  spaces = "    ";
}
if (freq >= 100 && freq <= 999)
{
  spaces = "    ";
}
if (freq >= 1000 && freq <= 9999)
{
  spaces = "   ";
}
if (freq >= 10000 && freq <= 99999)
{
  spaces = "   ";
}
tft.print(spaces);
delay(tft_delay);
tft.print(freq, 0);
delay(tft_delay);
tft.print(" Hz");
delay(tft_delay);
tft.setTextSize(status_font_size);
delay(tft_delay);
tft.setCursor(40, 90);
delay(tft_delay);
tft.print(wave_mode);
delay(tft_delay);
tft.print(" Wave");
delay(tft_delay);
tft.setCursor(20, 115);
spaces = " ";
if (current_step_size == 1)
{
  spaces = "  ";
}
if (current_step_size == 10)
{
  spaces = "  ";
}
if (current_step_size == 100)
{
  spaces = "  ";
```

```
  }
  if (current_step_size == 1000)
  {
    spaces = " ";
  }
  tft.print(spaces);
  delay(tft_delay);
  tft.print("Step Size: ");
  delay(tft_delay);
  tft.print(current_step_size, 0);
  tft.print(" Hz");
  tft.setTextSize(text_font_size); // Set the Font Size
  delay(tft_delay);
}
```

The read_encoder() function will read the current position of the rotary encoder. If the position has changed, we'll increase or decrease the frequency based on the direction the encoder was rotated and in increments based on the frequency step rate. This function will also set the AD9833 to the current frequency and update the frequency display on the TFT display.

```
void read_encoder()
{
  // Read the Encoder
  encoder_Pos = Enc.read() / 4;         // divide by 4 to match encoder detent
  if (encoder_Pos != old_encoder_Pos)  // If the Encoder has changed update freq
  {
    if (encoder_Pos > old_encoder_Pos) // If we're increasing frequency
    {
      if (freq + current_step_size >= max_freq) // Limit to top end of band
      {
        freq = max_freq;
      } else
      {
        freq = freq + current_step_size;
      }

    } else
    {
      if (freq - current_step_size <= min_freq) // We're decreasing frequency,
limit to minimum
      {
        freq = min_freq;
      } else
      {
        freq = freq - current_step_size;
```

```
      }
    }

    old_encoder_Pos = encoder_Pos;

    AD.setFrequency(MD_AD9833::CHAN_0, freq);

    tft.fillRect(0, 50, 159, 25, ST7735_BLUE); // Erase the lower portion of the
TFT display
    delay(tft_delay);
    tft.setTextSize(freq_font_size);
    delay(tft_delay);
    tft.setCursor(5, 50);
    delay(tft_delay);
    // Add spaces to center the frequency display
    spaces = «  «;
    if (freq < 100)
    {
      spaces = «    «;
    }
    if (freq >= 100 && freq <= 999)
    {
      spaces = «    «;
    }
    if (freq >= 1000 && freq <= 9999)
    {
      spaces = «   «;
    }
    if (freq >= 10000 && freq <= 99999)
    {
      spaces = «   «;
    }
#ifdef debug
    Serial.print(«Freq: «);
    Serial.println(freq);
#endif

    tft.print(spaces);
    delay(tft_delay);
    tft.print(freq, 0);
    delay(tft_delay);
    tft.print(« Hz»);
    delay(tft_delay);
  }
}
```

The read_pushbutton() function will check to see if the pushbutton switch on the rotary encoder has been pressed. If it has, it will first delay for the debounce delay period to allow the switch closure time to settle, then it will set the frequency step rate to next value in the step_size array. If the step size is already at the maximum step value, the step value will be reset to the lowest step value. The TFT display is then updated with the new step rate information.

```
void read_pushbutton()
{
  // Read the Encoder Pushbutton Switch
  encoder_Switch = digitalRead(encoder_Switch_Pin);
  if (encoder_Switch == LOW && millis() > delay_time) // Check to see if pressed
  {
    // if it's changed, toggle the step size but don't allow again for debounce
time
    delay_time = millis() + debounceInterval; // if it's changed, toggle the step
size but don't allow again for debounce time
    step_index = step_index + 1;
    if (step_index > 4) // We're at max step, reset to minimum step value
    {
      step_index = 0;
    }
    current_step_size = step_size[step_index];

#ifdef debug
    Serial.print("Step Size: ");
    Serial.println(current_step_size);
#endif

    tft.fillRect(0, 115, 159, 12, ST7735_BLUE); // Erase the lower portion of the
TFT display
    delay(tft_delay);
    tft.setTextSize(status_font_size);
    delay(tft_delay);
    tft.setCursor(20, 115);
    delay(tft_delay);
    spaces = " ";
    if (current_step_size == 1)
    {
      spaces = "  ";
    }
    if (current_step_size == 10)
    {
      spaces = "  ";
    }
    if (current_step_size == 100)
```

```
    {
      spaces = "  ";
    }
    if (current_step_size == 1000)
    {
      spaces = " ";
    }
    tft.print(spaces);
    delay(tft_delay);
    tft.print("Step Size: ");
    delay(tft_delay);
    tft.print(current_step_size, 0);
    tft.print(" Hz");
  }
}
```

The read_wave_type() function will read the SPDT Waveform Type switch and set the AD9833 Waveform Mode based on the switch position. A switch...case() statement is used to determine the Mode setting based on the switch position. In the center-off position, both digital I/O pins are HIGH and the Sine Wave Mode is selected. The AD9833 is set to the selected mode and the TFT display is updated with the new Mode information.

```
void read_wave_type()
{
  // Read the Wave selection switch and update the wave type if it has changed
  // Convert the Wave type switch input pins to a single binary value
  // Note, wave type cannot be 0 since the two I/O pins are attached to a SPDT
switch
  wave_type = digitalRead(Square_wave) + (digitalRead(Triangle_wave) * 2);

  if (wave_type != previous_type)
  {
    // Switch has changed, update the wave type
    switch (wave_type)
    {
      case 1:
        // Square Wave
        AD.setMode(MD_AD9833::MODE_SQUARE1);
        wave_mode = " Square";

#ifdef debug
        Serial.println("Set Wave Type to Square Wave");
#endif

        break;
```

```
        case 2:
          // Triangle Wave
          AD.setMode(MD_AD9833::MODE_TRIANGLE);
          wave_mode = "Triangle";

#ifdef debug
          Serial.println("Set Wave Type to Triangle Wave");
#endif

          break;

        default:
          // Sine Wave
          AD.setMode(MD_AD9833::MODE_SINE);
          wave_mode = "  Sine";

#ifdef debug
          Serial.println("Set Wave Type to Sine Wave");
#endif

          break;
      }
      previous_type = wave_type;
      // Update the display

      tft.fillRect(0, 90, 159, 20, ST7735_BLUE); // Erase the lower portion of the
TFT display
      delay(tft_delay);
      tft.setTextSize(status_font_size);
      delay(tft_delay);
      tft.setCursor(40, 90);
      delay(tft_delay);
      tft.print(wave_mode);
      delay(tft_delay);
      tft.print(" Wave");
      delay(tft_delay);
    }
  }
```

Once the Waveform Generator was finally up and running, an output amplifier circuit was added. I chose to use a Texas Instruments LM386 Audio Power Amplifier chip. While Texas Instruments has officially announced the end of life for the LM386 audio amplifier IC, it is still readily available from the usual online sources such as eBay, Amazon, AliExpress and others. An alternative amplifier IC would be the LM4862, capable of 675 mW of audio output. If you would prefer to use the LM4862 instead of the LM386, there are a number of simple amplifier designs you could use instead of the LM386 circuit used in this project.

I prefer the LM386 because it can operate from 4 to 12 volts instead of the 2.7 to 5 volt limitation of the LM4862, even though the LM4862 has nearly twice the power output of the LM386.

The LM386 is a single power supply voltage, self-contained audio power amplifier, with a selectable gain from 20 to 200, bandwidth of 300 kHz, and an output of 325 mW. I also chose the LM386 over a standard Op Amp because the LM386 does not require a feedback resistor or any other external components to operate at its default gain of 20. The 300 kHz bandwidth of the LM386 was more than adequate for what the design of the Waveform Generator called for. If you desire your Waveform Generator to operate up to the full 12.5 MHz bandwidth of the AD9833 DDS module, you'll need to replace the LM386 circuit with an op amp or similar amplifier circuit capable of handling the higher frequencies.

Once the LM386 was added to the breadboard, along with a 10-turn, 10-kΩ potentiometer on the input to the LM386 for level control, the project was soldered up on a 60 mm by 80 mm prototyping board and mounted in a Solarbotics MegaSAFE enclosure. The external switches, frequency control rotary encoder, and output jack were mounted to one of the enclosure's side panels.

Once everything was mounted in the enclosure, some final tests were run to see how everything performed. By using a DDS module to generate the waveforms (Figures 12.16 thru 12.18), you can see that the output waveforms were very crisp and distortion free across the entire 12.5 MHz operating range of the AD9833 DDS module (measured before the LM386 Amplifier). Due to the coupling capacitor on the LM386, the edges of the square wave on the output of the LM386 started to lose their sharpness around 50 kHz, well above the 20 kHz design criteria point desired when the circuit was designed.

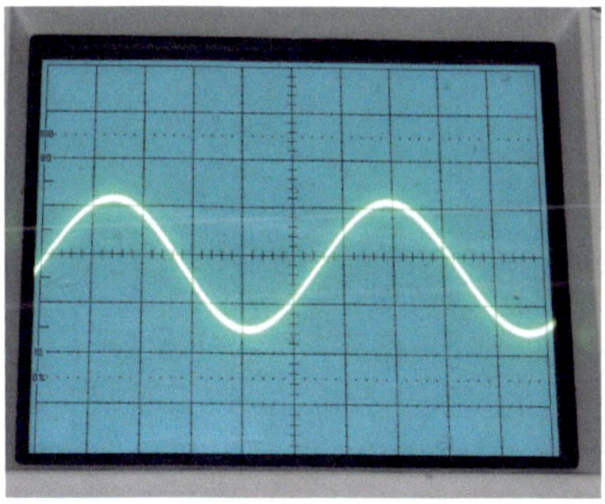

Figure 12.16: A 1 kHz sine wave generated by the AD9833 DDS.

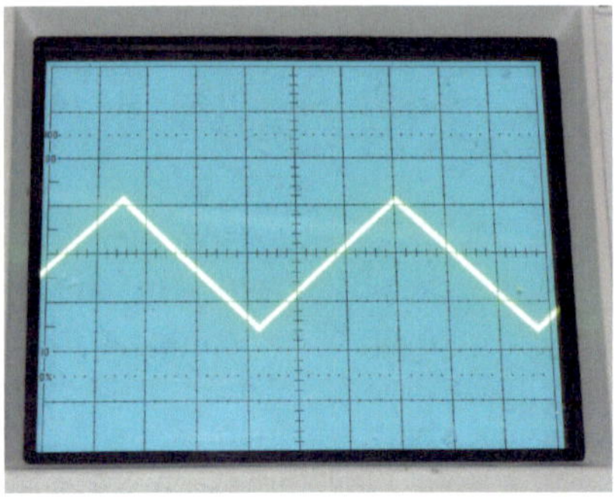

Figure 12.17: A 1 kHz triangle wave generated by the AD9833 DDS.

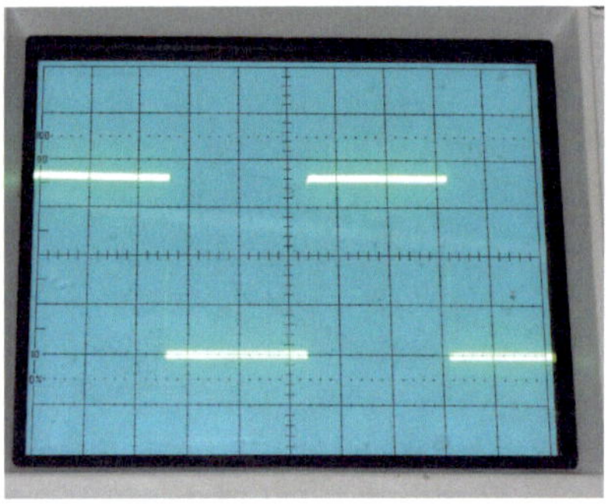

Figure 12.18: A 1 kHz square wave generated by the AD9833 DDS.

The real fun came when I hooked up a frequency counter to see how accurate the DDS really was while generating an 870-Hz audio signal (Figure 12.19). They aren't lying about the 0.1 Hz resolution. The frequency output is dead on accurate across the entire range. So, what I thought was going to be a simple waveform generator project has been transformed into a precision signal generator. You could actually use this circuit as the basis for a highly accurate VFO with a digital frequency display, although I would prefer to use the AD9850 DDS in an HF VFO project since it has a frequency range of up to 62.5 MHz.

Figure 12.19: AD9833 DDS set to 870 Hz.

Enhancement Ideas

Since this project ended up way beyond being just an audio frequency waveform generator, there's a lot you can do with it. In my version of the Waveform Generator, I did nothing with the phase shift features of the AD9833. You could add in a second rotary encoder to adjust the phase shift of the output waveform. You could add a multi-position rotary switch as a band switch, and use this as the basis for an HF VFO, or even a QRP transmitter. This is one of those projects that is only limited by your imagination. I started out being somewhat leery of the DDS modules, and now that I know how easy they really are to use, I can think of a whole list of projects I want to use with the DDS modules.

Chapter 13 • Project 7 — Auto Power On/Off

Figure 13.1 The finished Auto Power On/Off Project>

When we were brainstorming some project ideas for my first Arduino book, Tim Billingsley, KD5CKP (SK), suggested an Auto Power On/Off (APO) project similar to the APRS World APO3 and the West Mountain Radio ISOpwr+ units. This type of device is typically used in mobile operation to prevent the vehicle battery from being completely discharged when the vehicle's engine is not running, and the battery is not being charged. These devices turn on the DC power to your gear automatically when the vehicle's engine is running and the battery is being charged. These units will typically delay several minutes before turning the DC power off after the engine is shut off.

At that time, I did not see a need for this kind of project. After all, how much power does a mobile transceiver use when it's left on and just receiving, even if I did forget to turn it off? That was before I got a new car and added a second rig and an APRS transmitter (see the sidebar about APRS). Add to that the few weeks when the car just sat in the driveway, with the rigs and APRS forgetfully left powered on, and poof, dead battery. Ok, now I get it. I immediately bought and installed an APO3 Auto Power Off unit from APRS World to deal with that issue. When it came time to think up projects for this book, Tim's idea for the Arduino-powered APO project immediately came to mind.

Automatic Packet Reporting System (APRS)

The Automatic Packet Reporting System, also known as APRS, is an amateur radio-based tracking system developed by Bob Bruninga, WB4APR (SK), that is used for real-time tracking and digital communication of information. While commonly used by hams for vehicle tracking, APRS is capable of much more. APRS was designed to support a rapid and reliable exchange of information for local, tactical real-time information, events, or nets. APRS data is sent using standard 1200-baud, X.25 Packet Radio devices such as TNCs (Terminal Node Controllers) and can include GPS coordinates, weather station telemetry, text messages, announcements, queries, and other telemetry. APRS can even be used to send email to the Internet.

In the North America, 144.39 MHz is dedicated for APRS use throughout the continent (in Europe, the frequency is 144.800 MHz, in Australia, 144.175 MHz). While primarily intended for local, tactical use, through a network of packet digipeaters (digital packet radio repeaters), APRS data is retransmitted throughout an area, and can also be routed to the Internet through was is known as an IGate (Internet Gateway). IGates can be used to gather APRS data, and forward this information to the Internet APRS backbone, for use by websites such as www.aprs.fi, which can be used to track APRS users and view received APRS data, messages, and telemetry. For more information, please visit the APRS website at www.aprs.org.

The concept behind an APO unit is simple. When a car engine is running, the alternator is charging the battery, and the DC voltage is above the normal voltage of the battery alone. When the engine is not running, the DC voltage is whatever voltage the battery is providing. When the voltage is higher than a preset threshold, the APO will turn on the power to the load. When the engine is shut off and the voltage drops, after a pre-set time delay, the APO will shut off the power to the load. In the case of the APO3, it can handle a maximum load of 30 A, so our Arduino-powered version would need to be able to handle a similar load.

I initially designed the APO project with the intent of using a power MOSFET like the big boys do to switch the power to the load. Unfortunately, I had never worked with power MOSFETs before, and did not have a great deal of success trying to switch the amount of DC current needed to supply the load. After losing a few rounds to Ohm's Law and the embarrassing puffs of victory smoke, I decided to fall back and regroup. I'm sure the few remaining power MOSFETS in my parts bin were happy with that decision. I'm also sure that if I had spent more time working with power MOSFETs I could have eventually figured it out, but the idea behind these projects is simplicity and functionality. Since I wasn't comfortable switching the load with a power MOSFET, I fell back to the old reliable relay. Using a relay gives the added benefit of being able to select the maximum amount of current the APO project can supply to the load, simply by using a different relay.

Design Parameters

Starting out, I needed to know the turn-on and turn-off voltage thresholds I would want to use in this project. Using a multimeter, I measured the voltage on my car when the engine was running, and measured 13.7 V. The voltage dropped to 13.2 V when the engine was first shut off and drifted down to 12.8 V after a few minutes. Based on this, it was decided that the design criteria would be to turn the power on to the load 30 seconds after the voltage went higher than the turn-on threshold, and to turn the load off 10 minutes after the voltage went below the turn-off threshold. At the expiration of the turn-on or turn-off delay, the voltage thresholds would be checked again to verify that the voltage still met the criteria for the desired action. This would stop a turn-on if the engine was shut off again or stop a turn-off if the engine was restarted before the delay expired.
Additionally, the project would have a bypass switch that would provide power to the load, regardless of the input voltage. When the bypass switch is turned off, the APO will revert to whichever operating condition was called for based on the input voltage. An ST7735-type color TFT display will be used to display the status of the threshold voltage and delays.

The Project Design

Now that I had the design parameters and a concept for a design that I felt was viable, it was time to start designing the project in earnest. The first step in my build process is to draw up a block diagram of the project as shown in Figure 13.2, so that I will have a general idea of what parts will be needed, and what type of enclosure would be best suited to contain everything. Based on the design parameters and the block diagram, I chose to use an Arduino Nano mounted on a 60 mm by 80 mm piece of copper-clad perfboard mounted inside a Solarbotics MegaSAFE enclosure to allow room for all the components and to keep the project in a small protective box.

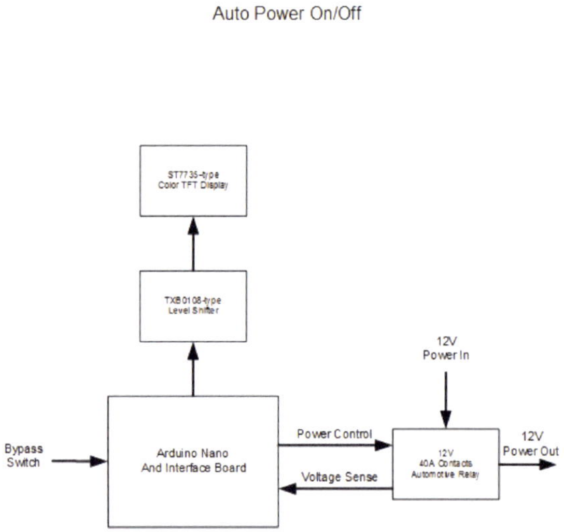

Figure 13.2: The Auto Power On/Off block diagram.

The Project Flowchart

After I have the basic design concept for a project, the next thing I do is prepare a quick flowchart of what the Arduino sketch needs to do. I use the flowchart as a form of a block diagram for the software. When combined with the hardware block diagram, I get a high-level overview of the project, both hardware and software. These two drawings allow me to cross-check my design and look for things I need in software that aren't planned for in the hardware, and vice versa. As you can see from the project flowchart (Figure 13.3), there's actually quite a lot going on in the Arduino sketch.

Next, we'll use the block diagram and the design parameters to build the project on a breadboard. Before I start the actual build, I'll usually sketch out a hand-drawn version of the schematic diagram to work from and keep track of any design changes that need to be made. Once I have a working hardware design, I stop to take the time to document the project. I do this for all the projects I build, not just for the ones I plan to be in a book. If I do the documenting while everything is fresh in my mind, it's a lot easier to make sure I have all the important things noted, and that the drawings are correct. Who knows, I may revisit this project later on down the road, and having these drawings to fall back on are a great help for remembering what I did and why I did it the way I did.

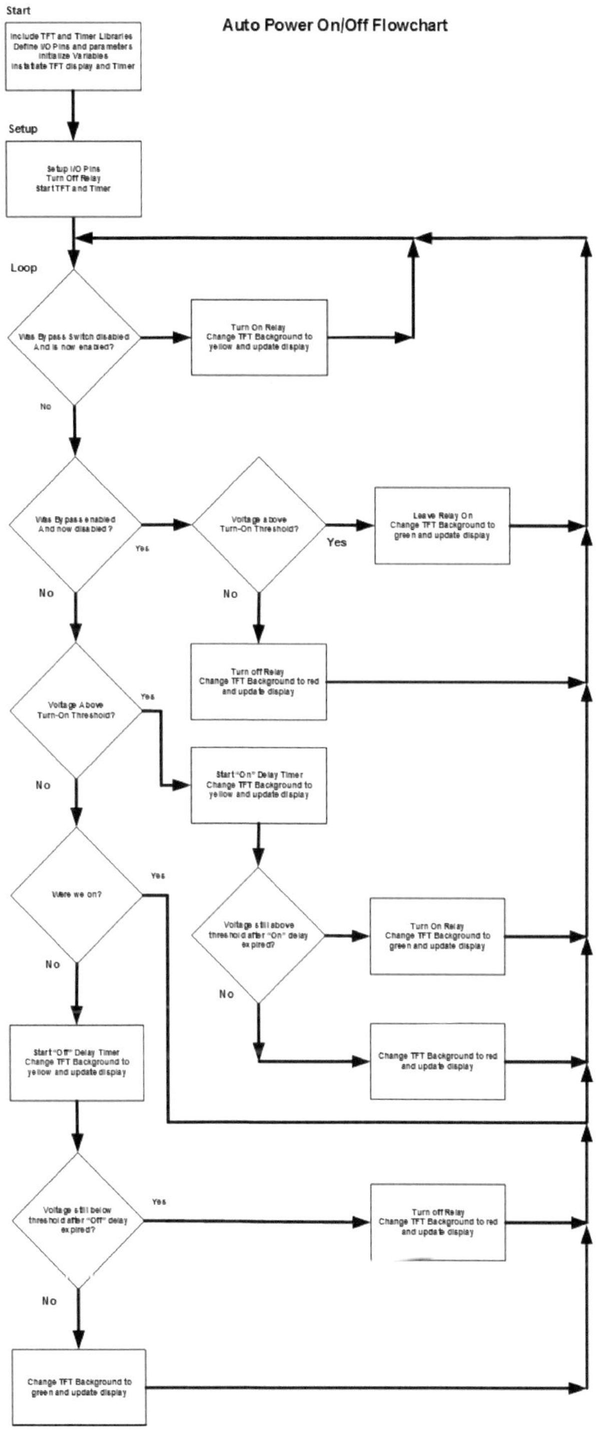

Figure 13.3: The Auto Power On/Off flowchart.

While the project is still in the development phase, I'll usually make all of my notes on the handwritten schematic. Until the project is finished, there's no real reason to take the time to create the permanent documentation. At this point, I will also take the time to write out a parts list of all the parts used in the project. This makes it a whole lot easier if I plan on building another one later on. I can gather up the parts I need, and order any that I don't have on hand, before I start on the project. Figures 13.4 and 13.5 show the finished schematic and parts list for the Auto Power On/Off project. One thing to note is that we use a voltage divider network to reduce the input and load voltages to a range between 0 and 5 volts to prevent damage to the Arduino's analog input pins.

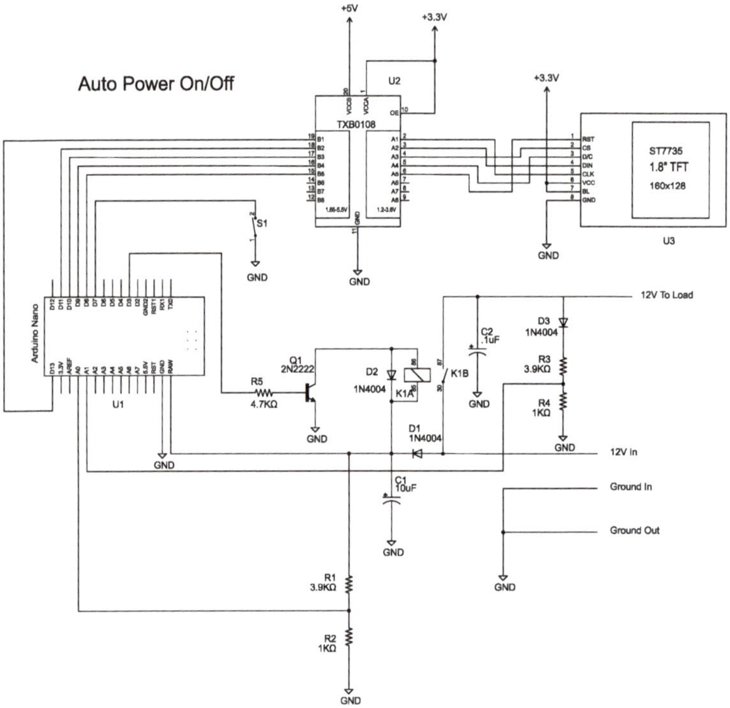

Figure 13.4: The Auto Power On/Off schematic.

Parts List
Auto Power On/Off

Part	**Value**
C1	0.1 µF, 50 V, ceramic capacitor
C2	10 µF, 25V, capacitor,
D1, D2, D3	1N4004 diode
K1	12 V coil, 30 A minimum contacts relay
Q1	2N2222 transistor
R1, R3	3.9 kΩ resistor, 1/8W
R2, R4	1 kΩ resistor, 1/8W
R5	4.7 kΩ resistor, 1/8W

S1	SPST switch
U1	Arduino Nano
U2	TXB0108 level shifter module
U3	ST7735 TFT display
Enclosure	Solarbotics Arduino MegaS.A.F.E

Figure 13.5: The Auto Power On/Off parts list.

From the now-updated handwritten schematic, the next step is to wire up the final version of the project. I chose to mount the project on the usual 60 mm by 80 mm piece of copper-clad perfboard that will mount inside the Solarbotics MegaSAFE with room left over for all of the external pieces of the project as shown in Figure 13.6. The Nano was mounted in a socket made from 2 pieces of 0.1-inch (2.54-mm) SIP (Single In-Line package) socket strips cut to size, and 0.1-inch DuPont style header pins were used for connecting to the external components. The relay I ended up choosing was a 12-volt automotive accessory relay with 40 A contacts that I got from a local auto parts store. I chose this relay because it had a convenient mounting flange that allowed me to mount the relay to the outside of the enclosure. I chose to have the relay on the outside of the enclosure to keep high current 12 volts away from the sensitive internal parts if the relay ever broke loose due to vibration. Using my handy American Wire Gauge (AWG) cheat sheet, I chose to use AWG #12 wire for the power and load wiring, as this would allow for a maximum of 41 A on the wiring itself. Since the relay contacts are 40 A, and I'm planning to have a 30-A maximum load, this fits the design parameters. Be sure to use the correct wire size needed for the current you plan to have your APO project switch. I used Anderson Powerpole® connectors for the DC power input and the Load output connections to keep everything consistent with the rest of my rig power connections.

Figure 13.6: The Auto Power On/Off circuit board assembly.

Creating the Sketch

All that's left now is to create the actual sketch we'll use to make the APO project work. While we will break the sketch down and examine it piece by piece to help you understand how it all works, the completed sketch and libraries for this project can be downloaded from www.kw5gp.com/Elektor.

To create the sketch, we will use the flowchart we created earlier to break the sketch down into smaller building blocks. This allows us to test each block as we go along. At the top of the sketch, you will find any general comments regarding the sketches, and any attributions that may be required by the licensing conditions of the various libraries and shared code that your sketch may use. Below that is where the actual sketch begins, usually starting with the library #include statements, variable definitions, and object assignments for any library components we will be using. In most, if not all of my sketches, you will also see a #define debug statement. I use this to enable the outputting of debugging information to the IDE Serial Monitor while creating and testing the sketch. When the sketch is complete, I comment out the debug definition statement, and the #ifdef...#endif blocks of debug code are ignored by the IDE when the sketch is compiled, and no output is sent to the Serial Monitor. Since this is a standalone project that will not need to communicate with the IDE, it makes sense to disable the debug output in the finished project.

All of the delay timing in the APO project is handled by the Timer.h library by J. Christensen, available at https://github.com/JChristensen/Timer. The Timer.h library is used to keep track of the various timing events that can occur while the APO sketch is running. The Timer library will generate timer interrupts when an event occurs, rather than having to wait in delay() loops. Because of all the possible interaction from external power threshold events, using delay() statements would not have worked in this project, because a threshold voltage change could occur while we're delaying and the sketch would miss it. When you use the delay() statement, it effectively pauses the sketch at that point and nothing else happens until the delay expires. The only viable alternative would be to have a series of manual counters running to keep track of time, and constantly check for things to occur. Using the Timer library saves us from having to do all that and takes care of keeping track of the various timing events for us.

To start out with the sketch, we include the required libraries and define the I/O pins and timer delay settings needed for the project.

```
#include <Adafruit_GFX.h>       // Core graphics library
#include <Adafruit_ST7735.h>    // Hardware-specific library
#include "Timer.h"              // Timer Library so we can do timed interrupts

#define TFT_CS      10          // Assign the TFT CS to pin 10
#define TFT_DC      9           // Assign the TFT DC to pin 9
#define TFT_RST     8           // Assign the TFT RST to pin 8

#define voltage_pin A0          // The input analog voltage pin
#define load_voltage_pin A1     // The output analog voltage pin
#define Relay 3                 // The Relay drive pin
#define Bypass_Switch 7         // The Bypass switch input pin
#define text_font_size 2        // The TFT text display font size
#define status_font_size 1      // The font size for the status line
#define tft_delay 10            // set the TFT command delay to 10ms
#define on_delay 30000      // On delay in milliseconds - Turn on after 30 seconds
```

```
#define off_delay 1200000  // Off delay in milliseconds - Turn off after 20
minutes
```

In the next part of the sketch, we declare and initialize the variables and constants we need to calculate the input voltage, set the delay times, and keep track of the current state of the APO. The turn-on delay is set to 30 seconds and the turn-off delay is set to 10 minutes. You may want to change these values to suit your needs.

```
float in_voltage;
float out_voltage;
float volt_drop_in = .775;      // The voltage drop across the 1N4004
float volt_drop_load = .616;    // The voltage drop across the 1N4004
float volt_threshold = 13.4;    // The turn-on threshold voltage
int volt_cal_in = 2520;         // Voltage calibration value for the MAP statement
int volt_cal_load = 2520;       // Voltage calibration value for the MAP statement
float previous_in_volts = 0;
float previous_out_volts = 0;
boolean volts_changed = false;

boolean current_state = false;  // The current power status
boolean bypass_Enabled = false; // Bypass active value
boolean timing = false;         // Variable to let us know that timing has started

boolean status_change = false;  // Powering On/Off delay in progress flag
long seconds = 0;               // The number of seconds since the timer started
```

Next, we need to set up the variables to keep track of the various timer events. The Timer library can keep track of multiple timing events simultaneously and assigns a process number to each event that we're monitoring. For this project, we'll have 3 separate timing events, in addition to the library's standard one second "tick" event, which it uses to keep track of the number of seconds that have passed.

```
int tickEvent;       // The process number for the one second Timer tick
int On_afterEvent;   // The process number for the On Delay Timer event
int Off_afterEvent;  // The process number for the Off Delay Timer event
int background_color = ST7735_BLUE;
String Status = "Off";
int spaces = 60;     // The number of spaces for the status line
```

Finally, we instantiate the TFT display and Timer objects before moving on to the setup() portion of the sketch.

```
        // Instantiate the TFT Display
Adafruit_ST7735 tft = Adafruit_ST7735(TFT_CS,  TFT_DC, TFT_RST);  // Instantiate
the TFT display instance as tft

Timer t;  // Instatiate Timer instance as t
```

Now that we've defined everything, we're ready to do the actual setup and initialization of the APO. The setup() loop runs only one time before passing program control on to the main loop() where the actual sketch execution occurs.

The main purpose of the setup() loop is to set up the Arduino's I/O pin modes, set the relay to its initial power off state, and enable the internal pull-up resistor for the bypass switch pin.

```
// initialize the I/O pins
  pinMode(Relay, OUTPUT);
  digitalWrite(Relay, LOW);              // Turn off the power
  pinMode(Bypass_Switch, INPUT_PULLUP);  // Enable the internal pullup resistor
for the Bypass switch pin
```

In the last portion of the setup() loop, we'll display a brief startup screen on the TFT display, before clearing the display, reading the current input voltage, and updating the display with the current voltage and status information as shown in Figure 13.7.

```
  // Display the splash screen
  tft.initR(INITR_18BLACKTAB);   // initialize a 1.8» TFT with ST7735S chip,
black tab
  delay(tft_delay);
  tft.fillScreen(background_color);
  delay(tft_delay);
  tft.setRotation(1);            // Set the screen rotation
  delay(tft_delay);
  tft.setTextWrap(false);        // Turn off Text Wrap
  delay(tft_delay);
  tft.setTextSize(text_font_size); // Set the Font Size
  delay(tft_delay);
  tft.setTextColor(ST7735_WHITE);  //Set the Text Color
  delay(tft_delay);
  tft.setCursor(50, 20);         //Set the Cursor and display the startup screen
  delay(tft_delay);
  tft.print(«KW5GP»);
  delay(tft_delay);
  tft.setTextSize(2);
  delay(tft_delay);
  tft.setCursor(40, 60);
  delay(tft_delay);
```

```
  tft.print(«Arduino»);
  delay(tft_delay);
  tft.setCursor(15, 80);
  delay(tft_delay);
  tft.print(«Auto On/Off»);
  delay(tft_delay);

  delay(5000);
  background_color = ST7735_RED;
  readVoltage();
  update_display();
```

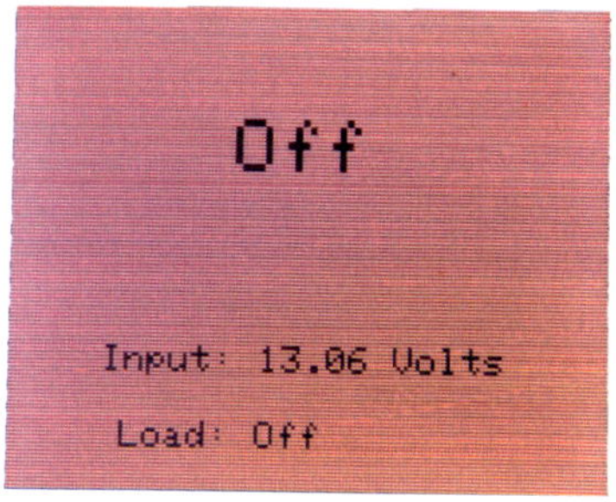

Figure 13.7: The Auto Power On/Off "Off" status display.

All of the real fun happens in the main sketch loop(). The main loop() is executed continuously until the Arduino is reset or powered off. The first thing the sketch does is check the status of the Bypass switch. If the Bypass switch is enabled, this overrides any timing events or conditions. The relay that switches the input power to the load is turned on, and we update the status on the TFT display as shown in Figure 13.8.

```
void loop()
{
  // Check for Bypass, overrides all
  if (digitalRead(Bypass_Switch) == LOW) // If Low, Bypass switch is on
  {
    if (!bypass_Enabled) // If we weren't already in bypass, turn on bypass mode
    {
      // Change the status to "Bypass"
      Enable_Relay();
      readVoltage();
      background_color = ST7735_YELLOW;
      spaces = 40;
      Status = "Bypass";
      update_display();
    }
    bypass_Enabled = true;  // Set the Bypass on flag
  }
```

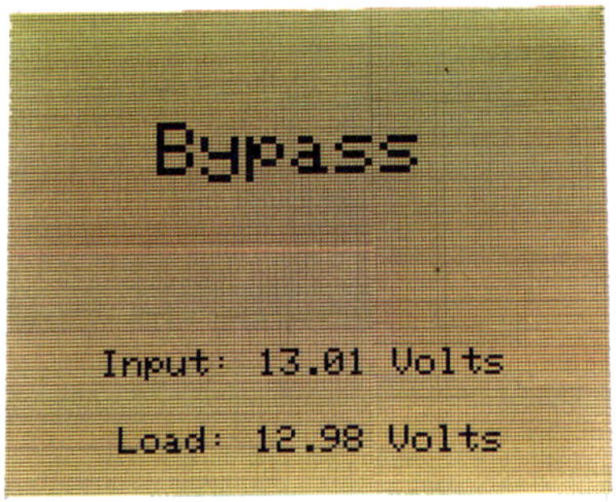

Figure 13.8: The Auto Power On/Off "Off Bypass" mode display.

If the Bypass switch is not turned on, we proceed with the remainder of the sketch. The first thing we do is to set the APO to the proper condition if the Bypass switch was turned on but is now off. This allows for the power to remain on if the threshold voltage is above the turn-on point, or to immediately turn off the power if the threshold voltage is below the turn-on point. The TFT status display is set to the appropriate background color, red for off, green for on, and the status information is updated.

```
else
  {
    // We're not in Bypass - see if it just changed status
    if (bypass_Enabled)
    {
      // Turn off the Relay only if the voltage is above threshold

      // We're not in Bypass - see if it just changed status
      bypass_Enabled = false; // Reset the Bypass flag
      // Change the status to On of Off depending on input voltage

      readVoltage();

      if (in_voltage >= volt_threshold)  // if voltage is above the turn-on
threshold we need to leave power on
      {
        current_state = true;
        Enable_Relay();
        background_color = ST7735_GREEN;
        Status = "On";
        spaces = 65;
      }
      else
      {
        current_state = false;
        Disable_Relay();
        background_color = ST7735_RED;
```

```
        Status = "Off";
        spaces = 60;
      }
      update_display();
    }
  }
```

The next portion of the sketch deals with the normal operation of the APO when not in Bypass mode. The sketch will read the input voltage and check to see if it is above the turn-on threshold. If we are above the threshold, but not already on, we update the TFT display to show a yellow background with the "On Delay" status information as shown in Figure 13.9.

```
if (!bypass_Enabled)
  {
    readVoltage(); // Read the input voltage
    // If we're not powered on, check to see if we need to be
    if (!current_state && (in_voltage >= volt_threshold) && (!status_change))
    {
      // Initiate Turn On Delay
      background_color = ST7735_YELLOW;
      Status = "On Delay";
      spaces = 35;
      update_display();
      status_change = true;
      On_afterEvent = t.after(on_delay, PowerOn); // Turn power on after start
delay
    }
```

Figure 13.9: The Auto Power On/Off "On Delay" status display.

When the 30 second turn-on timer expires, the Timer library calls the PowerOn() function shown below. This function will check the input voltage one last time before turning on the power to the load. If conditions are still ok for turning things on, we update the display to change to a green background and update the status information as shown in Figure 13.10. If the threshold voltage has dropped below the turn on point during the turn-on delay, the background color is changed to red, and the APO returns to the Power Off status display.

```
void PowerOn()  // Function to perform Power On operation after start delay
{
  // check to make sure we're still ready to power on

  readVoltage();
  if (in_voltage >= volt_threshold)  // We're still ok to power on
  {
    // Turn it on
    Enable_Relay();                   // Turn on the power
    background_color = ST7735_GREEN;
    Status = "On";
    spaces = 65;
    readVoltage();
    update_display();
    current_state = true;            // Set the current power status to On
  }
  else
  {
    // Not ok to power on
    current_state = false;
    background_color = ST7735_RED;
    Status = "Off";
    spaces = 60;
    update_display();
  }
  status_change = false;           // Turn off the Power On/Off delay status flag
}
```

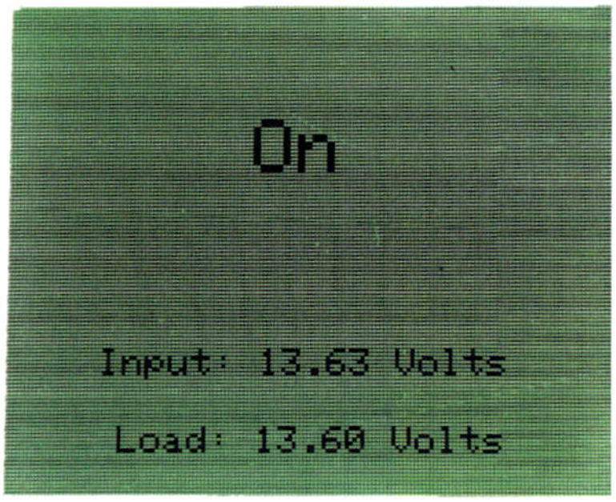

Figure 13.10: The Auto Power On/Off "On" status display.

The next portion of the main loop() deals with what to do when the voltage is below the turn-on threshold. If the APO is currently turned on, we change the display to a yellow background and display the Off Delay status information as shown in Figure 13.11.

```
// If we're powered on, check to see if we need to turn off
    // Turn off if below voltage threshold and we're powered on
    if (current_state && (in_voltage < volt_threshold) && (!status_change))
    {
      // initiate turn off delay
      status_change = true; // Turn on the flag indicating we're in a power on/
off delay
      background_color = ST7735_YELLOW;
      Status = "Off Delay";
      spaces = 25;
      update_display();
      Off_afterEvent = t.after(off_delay, PowerOff);  // Turn off after the turn-
off delay
      current_state = false;  // set the current power status flag to off
    }
  }
```

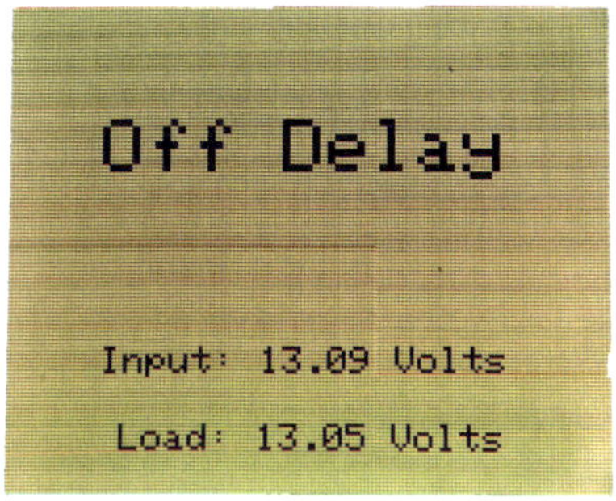

Figure 13.11: The Auto Power On/Off "Off Delay" status display.

When the turn-off delay timer expires, the Timer library will call the PowerOff() function shown below. This function will do one last check to verify that conditions are still valid to turn off power to the load. If the input voltage is still below the threshold voltage, we update the display to the Power Off status, and turn off the relay. If the input voltage is above the threshold value, the APO will remain on, and normal operation will continue.

```
void PowerOff()
{
  // check to make sure we're still ready to power off
  readVoltage();
  if (in_voltage < volt_threshold)  // We're still ok to power off
  {
    // Turn it off
    Disable_Relay(); // Turn off the power
    background_color = ST7735_RED;
    Status = "Off";
    spaces = 60;
```

```
    update_display();
    current_state = false;  // Set the current power status to Off
  }
  else
  {
    // Not ok to power off - leave power on
    background_color = ST7735_GREEN;
    Status = "On";
    spaces = 65;
    update_display();
    current_state = true;
  }
  status_change = false;  // Turn off the Power On/Off delay status flag
}
```

The last portion of the main loop shown below will read and update the displayed voltage and also perform the timer update() function that is required by the Timer library to update its internal timers used to keep track of timer events.

```
// Read the voltages and update display if they change more than .1V
  check_voltages();
  if (volts_changed)
  {
    update_display();
  }

  t.update();  // Update the Interrupt handler - required for the Timer library
to function
```

The update_display() function is used to set the appropriate display background color and display the current status information and voltages.

```
void update_display()  // Updates the TFT display
{
  tft.fillScreen(background_color);
  delay(tft_delay);
  tft.setTextColor(ST7735_BLACK); //Set the Text Color
  delay(tft_delay);
  tft.setTextSize(text_font_size);
  delay(tft_delay);
  tft.setCursor(spaces, 30);
  delay(tft_delay);
  tft.print(Status);
  delay(tft_delay);
  tft.setTextSize(status_font_size);
  delay(tft_delay);
```

```
  tft.setCursor(25, 90);
  delay(tft_delay);
  tft.print("Input: ");
  delay(tft_delay);
  tft.print(in_voltage, 2);
  delay(tft_delay);
  tft.print(" Volts");
  delay(tft_delay);
  tft.setCursor(30, 110);
  delay(tft_delay);
  tft.print("Load: ");
  delay(tft_delay);

  if (out_voltage > 1)
  {
    tft.print(out_voltage, 2);
    delay(tft_delay);
    tft.print(" Volts");
  }
  else
  {
    tft.print("Off");
  }

  delay(tft_delay);
  tft.setTextSize(text_font_size); // Set the Font Size
  delay(tft_delay);

  previous_in_volts = in_voltage;
  previous_out_volts = out_voltage;
}
```

The readVoltage() function will read the input and load voltages. The actual value is mapped to a corrected value because we had to use voltage divider circuits to reduce these voltages to a value from 0 to something less than 5 Volts to prevent damage to the Arduino's analog input pins.

```
void readVoltage()  // Function to read the input voltage
{
  float input_volts = analogRead(voltage_pin);
  float load_volts = analogRead(load_voltage_pin);

  in_voltage = map(input_volts, 0, 1023, 0, volt_cal_in); // Map the A/D value to
a voltage in millivolts
  in_voltage = (in_voltage / 100) + volt_drop_in; // Convert to volts and
subtract the input circuit voltage loss
```

```
  out_voltage = map(load_volts, 0, 1023, 0, volt_cal_load); // Map the A/D value
to a voltage in millivolts
  out_voltage = (out_voltage / 100) + volt_drop_load; // Convert to volts and
subtract the output circuit voltage loss
}
```

The Tick() function is used exclusively for debugging and will display a one second timer "tick" on the IDE's Serial Monitor when debug mode is enabled.

```
void Tick() // Provides a one second Timer tick
{
#ifdef debug
  seconds = seconds + 1 ; // increment the seconds counter
  readVoltage();  // Read the input voltage
  Serial.print ("   Seconds: "); // Print the Seconds and voltage
  Serial.print (seconds);
  Serial.print ("   Input Volts: ");
  Serial.print(in_voltage);
  Serial.print ("   Load Volts: ");
  Serial.println(out_voltage);
#endif
}
```

The Enable_Relay() function is used to turn on the relay, and the Disable_Relay() function is used to turn the relay off.

```
void Enable_Relay()
{
  digitalWrite(Relay, HIGH); // Turn on the power
  delay(1000);
}

void Disable_Relay()
{
  digitalWrite(Relay, LOW); // Turn on the power
  delay(1000);
}
```

The check_voltages() function will read the voltages and turn on the volts_changed flag to indicate that one of the voltages has changed more than 0.1 Volts and will be used to determine if we need to update the display or not.

```
void check_voltages()
{
  // Reads the voltages and turns on volts_changed flag
  readVoltage();
```

```
    // Update display if voltage changes more than .1V
    if (abs(in_voltage - previous_in_volts) >= .1 || abs(out_voltage - previous_
  out_volts) >= .1)
    {
      volts_changed = true;
    }
    else
    {
      volts_changed = false;
    }
  }
```

Enhancement Ideas

As I mentioned earlier, all of the projects in this book are complete and fully functional as they are. However, as a general rule, I like to stop a bit short on each project, allowing room for you to add your own personal touches to the finished project. With each project, I typically include my ideas for modifications and enhancements that I would want to add to each project, as a way of helping you with ideas for creating your own personalized version of the project and furthering your Arduino skills. The goal of this is to encourage you to go beyond what is presented in this book, and to help you as you move forward and create your own Arduino projects.

For the APO project, there are a couple of changes and enhancements that I would recommend. While my own initial attempts to build the APO project using a power MOSFET instead of a relay did not meet with much success, I still like the idea of using the smaller power MOSFET and heat sink design and having everything mounted inside the enclosure. I'm reasonably certain that the MOSFET overheating issues that I ran into during my early tests were related to not having the correct gate bias on the IRF9540 P-Channel MOSFET that I was using in my test configuration. As strange as it may sound, I actually do go back and review some of the not-so-successful projects I build to figure out exactly why something did the unexpected or didn't work as planned. The power MOSFET issue is one I definitely plan to revisit at a later date.

Another enhancement would be to replace the larger TFT display with a smaller 128x64 Organic LED (OLED) display to provide real-time status and input voltage information. You could also add over-voltage and under-voltage checking and protection to your APO project, as a way to further protect your mobile equipment.

Chapter 14 • Project 8 — Bluetooth CW Keyer

Figure 14.1: The finished Bluetooth CW Keyer Master unit.

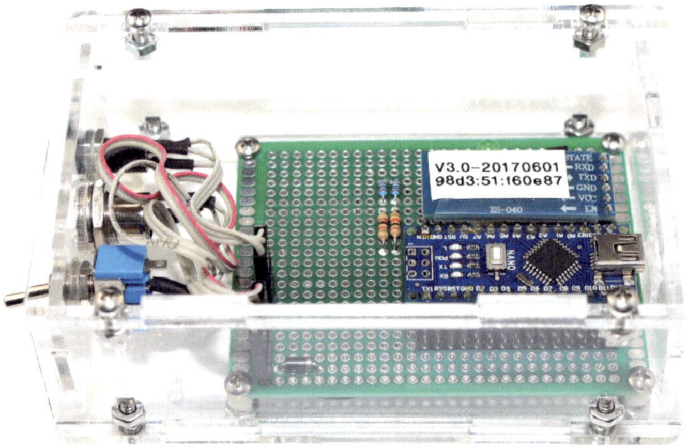

Figure 14.2: The finished Bluetooth CW Keyer Remote unit.

Ever since I started working with the Arduino, I've wanted to create some projects that use Bluetooth technology. One of the primary reasons I enjoy working with the Arduino and other microcontrollers is for the learning experience of working with new technologies and new components. Call me strange, but I enjoy figuring this sort of stuff out, simplifying and taking the mystery out of using these devices and technologies. Bluetooth technology fits squarely in that area for me. I've used Bluetooth headsets and other Bluetooth devices for years, but never really took the time to learn and understand the underlying technology. Now that the Arduino and other microcontrollers such as the Arduino R4, ESP32, and Raspberry Pi Pico, among others, have built-in Bluetooth, it's pretty clear that the time has come to start getting up to speed on Bluetooth technology.

To add some confusion to the process, it seems like the Bluetooth modules for the Arduino keep changing and evolving. The command set for the early HC-05 type Bluetooth modules is different than the command set used by the same modules available today. Much of the online documentation for the HC-05 Bluetooth module commonly used with the Arduino doesn't cover the newer versions and many of the commands either don't work as expected, or just aren't there anymore. The hardware design for the modules has changed as well, and it took quite some time to come up with a simple and functional ham radio-related project involving Bluetooth.

In actuality, I can't claim credit for this project idea. That credit belongs to ARRL CEO David Minster. I always enjoy visiting with the ARRL staff at the various hamfests and conventions I attend throughout the year. During a chat with Dave at the Orlando Hamcation earlier this year, he asked if I would be interested in building a special project for ARRL. He also mentioned that he'd asked several others, but they had as yet been unable to come up with what he was looking for. Special project? Arduino? Difficult or can't be done project? He managed to peg my curiosity, passion, and challenge meters at full scale all in a single sentence. How could I resist that? Challenge accepted, let the fun begin.

Dave quickly laid out the concept he was hoping to achieve as part of a theoretical "ham shack of the future" vision he was hoping to make a reality. In this particular case, he wanted a wireless CW keyer that would connect to the Flex transceiver at the W1AW ham shack at ARRL Headquarters. So, in short, he wanted a wireless CW Keyer for a wireless ham radio, or a wireless wireless CW keyer.

His vision kicked my brain into high gear, and I immediately started fleshing out a concept while gathering in the rest of the design criteria. The end goal would be to plug a standard set of CW keyer paddles into the remote unit, and the keyer contact closures would be relayed wirelessly to the Master unit connected to the Flex transceiver. Only the keyer contact closures would need to be sent as the Flex transceiver has a built-in CW keyer we'd be using.

My initial concept was to use an Arduino UNO or Nano and standard HC-05 type Bluetooth modules similar to the one shown in Figure 14.3 and 14.4 to create the link. Once the link was established, the remote would send ASCII characters representing the state of the CW paddle contacts to the Master unit that would translate the ASCII characters back into the keyer contact closures and drive a "dit paddle" and a "dah paddle" pair of relays that would then be connected to the CW keyer input on the Flex transceiver. Simple, low parts count, quickly constructed, that's how I like my Arduino projects to be. But could it really be that simple and actually work like he wanted? This is actually an important concept that needs to be highlighted here. When it comes to the Arduino, don't discount the simple "Bull in a China Shop" direct approach. All too often I run into problems with my Arduino designs when I overthink or overcomplicate the proposed solution. When it comes to the Arduino, keeping things simple and basic often gives you exactly what you're looking for without having to overbuild or make the code so complex that it becomes unworkable. Time and time again, I've fallen back to the simple solution to get the job done. As proof, that design philosophy is exactly what made this project work, a very simple and basic approach.

Figure 14.3: Top view of the HC-05 Bluetooth module.

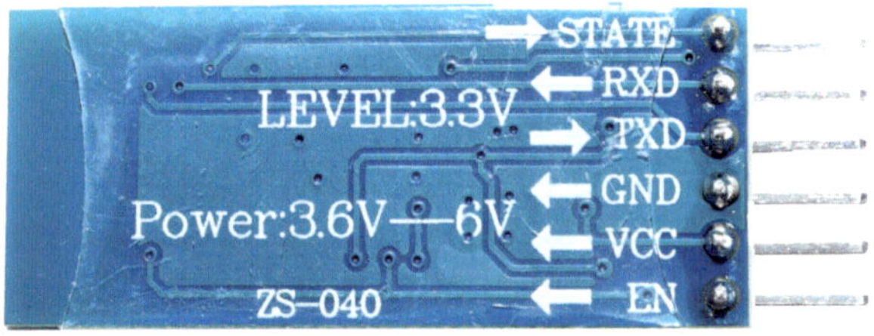

Figure 14.4: Bottom view of the HC-05 Bluetooth Module.

As is typical with a wild and crazy idea such as this, the further we moved along into the design concept process, the more doubts began to creep in. First, could I even get two of the current HC-05 Bluetooth modules to accept configuration commands, link up, and actually pass data between the two?

The current HC-05 module design has evolved from using a KEY pin input to now using an Enable (EN) input pin or the pushbutton switch on the board to switch between Command and Data mode. However, the sequence to get to the HC-05 Command mode is a bit tricky. During power-up, you can either ground the EN pin on the HC-05 module or hold the pushbutton down. Since the operation of this project was supposed to be seamless and not needing operator intervention, manually pressing the pushbutton switch would be out of the question. We'd need to connect the EN pin to a digital I/O pin on the Arduino and control it that way, except we also needed a way to turn the power on and off to the HC-05 module. However, we can't directly connect the Arduino's digital I/O pin to the HC-05 module's EN pin either.

It turns out that the HC-05 is a bit of a strange bird when it comes to power and the voltage applied to the HC-05's input pins. The HC-05 accepts from 3.6 to 6 volts on the power pin, however the input pins to the HC-05 can only accept a maximum of 3.3 volts. While we could use a level shifter module like the TXB0108 module, we really only need to shift the level on two pins on the HC-05 module, the EN (Enable) and RXD (Receive Data) pins. It seemed a bit of a waste to use an 8-channel TXB0108 level shifter module for just two pins. You could use a 4-channel TXB0104-type level shifter module instead of the TXB0108, but you'd still be wasting two channels. So, for this project, we'll use a simple resistive divider for the EN and RXD pins. The good news is that since the TXD and STATE pins are 3.3 volt output pins, we don't need to shift the level on those pins, they'll work fine just as they are.

The solution to the power on/off issue for the HC-05 module was to use a 2N2222 NPN transistor as a switch and have it control the power on and off to the HC-05 V_{CC} pin. Ok, so now we can control the HC-05 power and EN data pin to switch between Command and Data Mode, and we have safe input pin voltages that won't damage the HC-05 module. On to the next step.

But we're not out of the design issue woods yet. The HC-05 Bluetooth modules communicate using Bluetooth Serial Port Profile (SPP), which is vastly different that the HID (Human Interface Device) Bluetooth Profile that's used with Bluetooth headsets, keyboards, mice, etc. In SPP mode, the linked Bluetooth devices can send and receive serial data just as if there were wired connections between the modules and nothing else.

However, we're not trying to send serial data with this project, we're trying to send the CW keyer paddle switch contact closure information. How do we get the HC-05 to send that kind of information? What if we did send serial data and sent something like an ASCII "0" to indicate no paddle contacts are closed or when the paddles are released, an ASCII "1" if the dit paddle contact is closed, an ASCII "2" is the dah paddle contact is closed. When both paddles are released, we'd send the ASCII "0" character. This allows the iambic keying features of the Flex transceiver's built-in CW keyer to operate as if the keyer was hard-wired to it and, in theory at least, allow the duration of the keyer contact closures to be faithfully replicated on the other end.

Now the next set of design questions come into play. These questions would be a bit harder to answer without hands-on measurements and actual operation. Can the Bluetooth wireless serial link running at 38,400 baud send the keyer contact closure information fast enough so that a fast CW operator won't overrun the link speed? How badly would the overall latency from keyer contact closure time to actually keying the transceiver impact the whole process? Would the whole project appear seamless to the operator and perform as if the keyer was wired directly to the transceiver? These were unknowns that could only be determined once I'd built a working prototype.

Figure 14.5 and 14.6 shows the block diagrams for the Bluetooth CW Keyer Master and Remote units. There really isn't a lot to the hardware needed for this project. In fact, with just a few minor differences, the designs for the Master and Remote units are nearly identical. On the Remote unit, we'll use an Arduino Nano and an HC-05 Bluetooth module to read the keyer paddle contacts and send that information to the Master unit. The Master unit will also use an Arduino Nano and an HC-05 module to receive the keying information and drive a pair of reed relays connected to the CW Key input of the Flex transceiver.

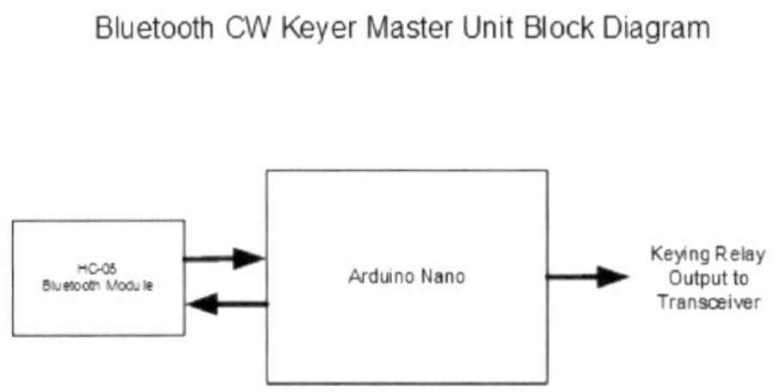

Figure 14.5: The Bluetooth CW Keyer Master Unit Block Diagram.

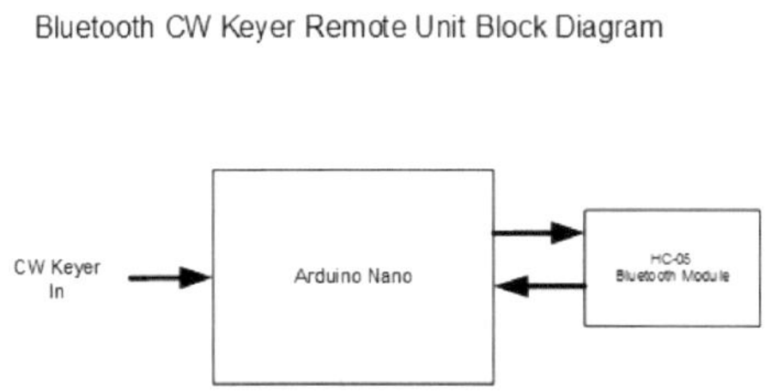

Figure 14.6 The Bluetooth CW Keyer Remote Unit Block Diagram.

The actual build for the Master and Remote are the typical "handful of wires" Arduino projects. Figure 14.7 shows the schematic diagram and Figure 14.8 shows the Parts List for the Master unit and Figure 14.9 shows the schematic diagram and Figure 14.10 shows the Parts List for the Remote unit. The Master and Remote units are nearly identical in terms of construction. The main differences between the two is the Master unit has a pair of reed relays used to provide the keying circuit contact closures to the transmitter, and the remote has the CW keyer paddles connected to a pair of digital I/O pins on the Arduino. The sketch for the two units will use virtually the same basic sketch framework, with the biggest differences being in the main sketch loop() as you'll see in a bit.

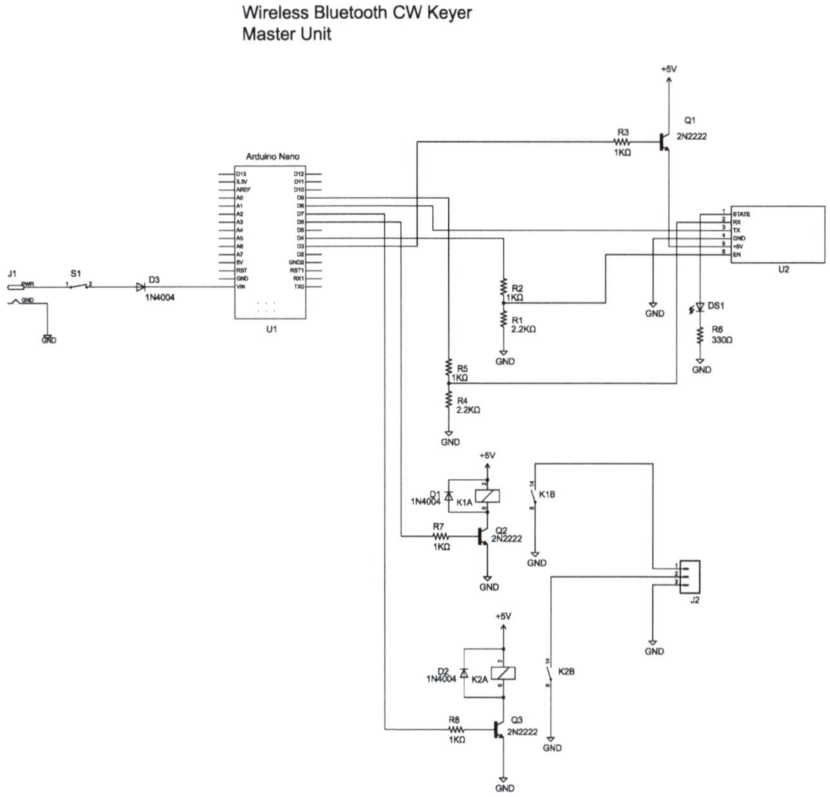

Figure 14.7: The Bluetooth CW Keyer Master Unit schematic.

Parts List
Bluetooth CW Keyer – Master Unit

Part	**Value**
D1, D2, D3	1N4004 diode
DS1	5 mm green LED
J1	DC power jack
J2	Mini stereo jack
K1, K2	5 V reed relay
Q1, Q2, Q3	2N2222A transistor
R1, R4	2.2 kΩ resistor, 1/8W
R2, R3, R5, R7, R8	1 kΩ resistor, 1/8W
R6	330 Ω resistor, 1/8W
S1	SPST switch
U1	Arduino Nano
U2	HC-05 ZS-040 Bluetooth module
Enclosure	Solarbotics Arduino S.A.F.E

Figure 14.8: The Bluetooth CW Keyer Master Unit parts list.

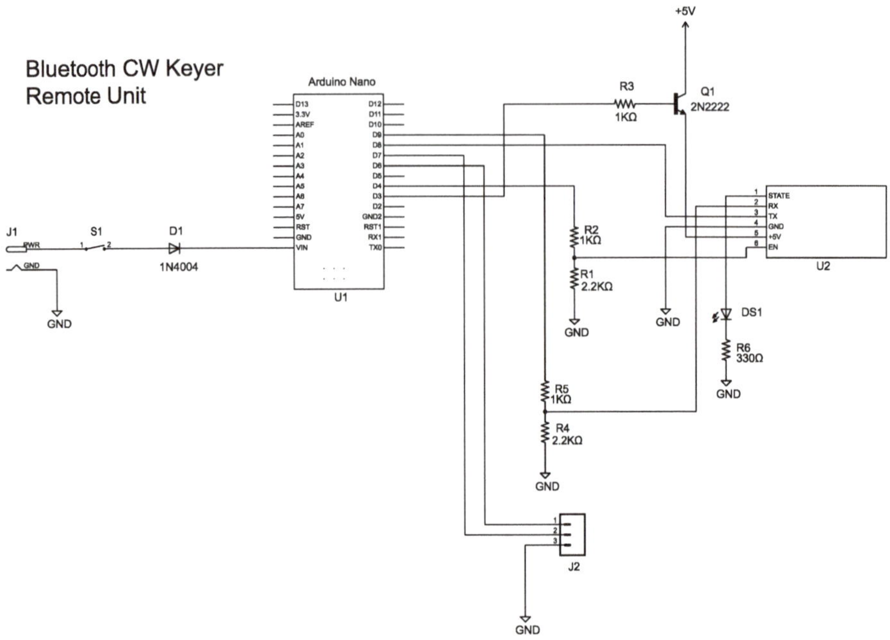

Figure 14.9: The Bluetooth CW Keyer Remote Unit schematic.

Parts List
Bluetooth CW Keyer – Slave Unit

Part	**Value**
D1	1N4004 diode
DS1	5 mm green LED
J1	DC Power Jack
J2	mini stereo jack
Q1	2N2222A transistor
R1, R4	2.2 kΩ resistor, 1/8W
R2, R3, R5	1 kΩ Resistor, 1/8W
R6	330 Ω resistor, 1/8W
S1	SPST switch
U1	Arduino Nano
U2	HC-05 ZS-040 Bluetooth module
Enclosure	Solarbotics Arduino S.A.F.E

Figure 14.10: The Bluetooth CW Remote Unit Parts List.

Figure 14.11 shows the circuit board for the Bluetooth CW Keyer Master Unit and Figure 14.12 shows the circuit board for the Remote Unit. The Arduino Nano and the HC-05 module are mounted to the board using Dupont-style 2.54mm spacing header sockets to allow for easy removal. The green LED on each of the boards is used to display the status of the HC-05's STATE pin, which will be lit when the Bluetooth link has been established.

The HC-05 module used in this project is the ZS-040 version and the firmware is the latest version available, V3.0-20170601, although the version of the HC-05's software shouldn't be a concern as long as the modules on either side of the link are using the same firmware version, and I'm not sure that would even be an issue, but I didn't want to push my luck at this point. You'll notice that the Remote Unit's HC-05 module's device address is labeled as 98d3:51:f60e87. This address is used by the Master unit to configure the link to "bind" to the specified remote unit address. This address is unique to each HC-05 module, and you'll need to place your remote's address in the sketch bind_addr definitions in the Master unit sketch. If you compile the sketches with the debug mode on, the addressing information will be included in the debug output.

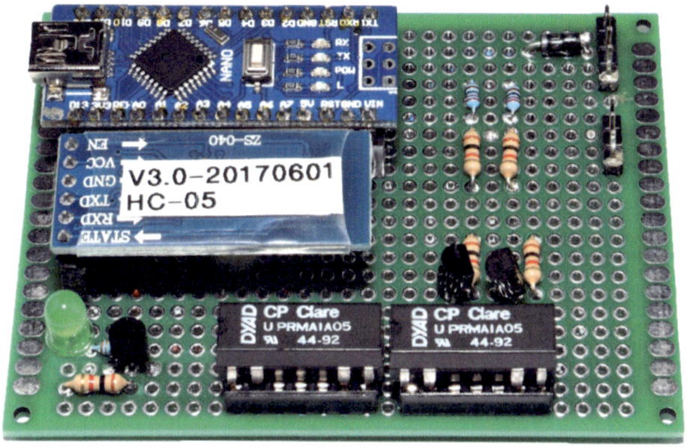

Figure 14.11: The Bluetooth CW Keyer Master Unit circuit board.

Figure 14.12: The Bluetooth CW Keyer Remote Unit circuit.

The Master Unit Flowchart

Let's go through the flowchart and sketch separately for each unit. Figure 14.13 shows the flowchart for the Bluetooth CW Keyer Master unit. The flowcharts for these projects are pretty straightforward. When you stop and look at what the units are actually doing, they're doing little more than sending three different ASCII characters based on the CW keyer paddle contact status across a wireless serial link, and then turning two relays on and off based on the ASCII characters received.

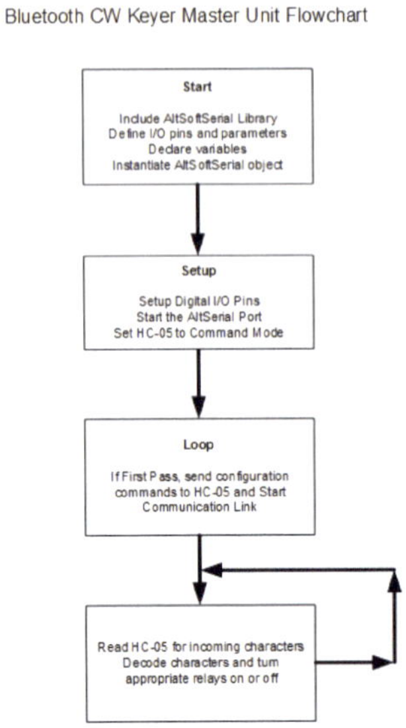

Figure 14.13: The Bluetooth CW Keyer Master Unit flowchart.

The Master Unit Sketch

Because the HC-05 module uses a TTL Serial connection, we'll need to use the AltSoftSerial Library to create a second serial port on pins 8 and 9 of the Arduino Nano. The Arduino UNO and Nano only have one hardware USB/Serial port on the board for sketch uploading and for the Serial Monitor used in debugging. For additional serial ports, you'll either need to use an external UART module or one of the Software Serial libraries to create a software serial port. I prefer to use the AltSoftSerial Library as it has better character buffering capability than the built-in Arduino Software Serial library. You'll also note that we don't use a library for the HC-05 Bluetooth module. I did manage to find a library for the HC-05, but without a lot more testing and experimentation with it, I don't feel comfortable using it and prefer to communicate directly with the HC-05 modules without using a library for the time being.

```
#include <AltSoftSerial.h>  // Include the software serial library
```

Next, we'll define the Serial and Bluetooth baud rates, along with the I/O pin definitions and the delay time between Bluetooth commands during startup. Theoretically you can change these baud rates, but these speeds seemed to work best in my testing of this project.

```
#define serial_baud 38400       // Set the Serial Monitor baud rate
#define bt_baud 38400           // Set the Bluetooth module baud rate
#define BT_PWR 3                // Define the Bluetooth module Power Control Pin
#define BT_EN 4                 // Define the Bluetooth module Enable Pin
#define Dit_pin 6               // Define the Dit Relay drive pin
#define Dah_pin 7               // Define the Dah Relay drive pin
#define command_delay 1000      // Delay between the Bluetooth commands
```

This is a really cool sketch in that it doesn't need many variables at all. In fact, the sketch for the Master unit only uses four variables.

```
bool run_once = true;
String rx_line;
bool receiving;
char c;
```

Finally, we'll instantiate the AltSoftSerial object and move on to the setup() portion of the sketch.

```
AltSoftSerial altSerial;  // Instantiate the AltSoftSerial instance
```

In the setup() portion of the sketch, we start out by setting the pin modes for the relay and Bluetooth module pins, then we ensure that the relays are turned off.

```
void setup()
{
  pinMode(Dit_pin, OUTPUT);   // Setup the Dit Relay pin for output
  pinMode(Dah_pin, OUTPUT);   // Setup the Dah Relay pin for output
  pinMode(BT_PWR, OUTPUT);    // Setup the Bluetooth module Power Control pin for
output
  pinMode(BT_EN, OUTPUT);     // Setup the Bluetooth Enable pin for output

  digitalWrite(Dit_pin, LOW); // Turn off both the Dit and Dah relays
  digitalWrite(Dah_pin, LOW);
```

Next, we start the AltSerial port at 38,400 baud, set the HC-05 module EN pin HIGH, and turn on the power to the HC-05 to put it into Command mode. We'll then read any responses received from the Bluetooth module. These responses would be displayed if the debug mode was enabled. In normal operation, we can ignore any response from the HC-05 module and assume that everything runs ok.

```
altSerial.begin(bt_baud);      // Start the Bluetooth module serial connection
  delay(command_delay);

  // Turn on Bluetooth module PWR and EN pins
  digitalWrite(BT_EN, HIGH);   // Turn on the Enable pin
  delay(command_delay);

  digitalWrite(BT_PWR, HIGH);  // Turn on the power to the Bluetooth module
  delay(command_delay);

  read_bt();                   // Read any responses from the Bluetooth module
  delay(command_delay);
}
```

In the main sketch loop(), on the first pass through the loop, we'll turn off the dit and dah relays and send a series of AT-type commands to configure the HC-05 Bluetooth module. First, we'll set the HC-05 Name to HC-05, the Role to Master, the Connect Mode to Any, and the UART (The Arduino-connected Software Serial Port) baud rate to 38,400. Finally, we set the EN pin logic LOW and preform a reset on the Bluetooth module to switch it into Data mode to establish the link to the remote unit. When the Bluetooth link is established, the STATE pin on the HC-05 module goes HIGH and turns on the green LED on the circuit board. Note that in between each command we add a brief delay to allow time for the Bluetooth module to send any command replies. These replies would be displayed on the Serial Monitor if debug mode was enabled.

```
if (run_once) // On the first pass, configure the Bluetooth module
  {
    digitalWrite(Dit_pin, LOW);          // Turn off the dit and dah relays
    digitalWrite(Dah_pin, LOW);

    delay(command_delay);
    altSerial.println("AT+NAME=HC-05"); // Send the Set Name command
    delay(command_delay);       // Read any responses from the Bluetooth module

    read_bt();                  // Read any responses from the Bluetooth module
    delay(command_delay);

    altSerial.println("AT+ROLE=1"); // Send the Master Role command
    delay(command_delay);
    read_bt();                  // Read any responses from the Bluetooth module

    altSerial.println("AT+CMODE=1");  // Send the Connect to Any command
    delay(command_delay);
    read_bt();                  // Read any responses from the Bluetooth module

    altSerial.println("AT+UART=38400,0,0"); // Send the Baud Rate command
    delay(command_delay);
    read_bt();                  // Read any responses from the Bluetooth module

    digitalWrite(BT_EN, LOW);  // Turn off the Enable pin
    delay(1000);

    // Start the Bluetooth module communication link
    altSerial.println("AT+RESET");  // Send the Bluetooth module Reset command
    delay(command_delay);
    read_bt();          // Read any responses from the Bluetooth module
    delay(command_delay);
    read_bt();          // Read any responses from the Bluetooth module
    flush_bt();         // Flush any additional responses
```

```
    run_once = false;
  }
```

Once all of the first pass Bluetooth configuration is complete and the link is established, we'll call the read_bt() function to check and process any incoming characters on the link.

```
read_bt();  // Read any responses from the Bluetooth module
```

The read_bt() function is used to check the Bluetooth module for any incoming characters and a switch...case() statement is used to process the incoming ASCII character and turn the proper relays on or off.

```
void read_bt()  // Function to read the Bluetooth module responses
{
  rx_line = "";
  do
  {
    if (altSerial.available())
    {
      receiving = true;
      c = altSerial.read();    // Get the incoming character

      if (!run_once)           // Process the incoming character only if we're not
doing configuration
      {
        switch (c)
        {
          case '0': // No CW paddle pressed - turn off both relays
            digitalWrite(Dit_pin, LOW);
            digitalWrite(Dah_pin, LOW);
            break;

          case '1': // Dit paddle press received - turn on dit relay
            digitalWrite(Dit_pin, HIGH);
            break;

          case '2': // Dah paddle press received - turn on dah relay
            digitalWrite(Dah_pin, HIGH);
            break;
        }
      }
    }
  } while (altSerial.available());
}
```

The flush_bt() function is used as a "housekeeping" type function to clear the HC-05's receive buffer so we don't have erroneous data in the buffer while starting up.

Next up, we'll go through the Remote Unit's Flowchart and Sketch. Figure 14.14 shows the Flowchart for the Bluetooth CW Keyer Remote unit. As you can tell, the flowchart for the Remote unit is nearly the same as the one for the Master unit. The main difference is that the Remote unit will read the CW paddle switch contacts on the Arduino's I/O pins, convert those contact closures to ASCII characters that represent the switch closures and then send those serial characters across the wireless link to the Master unit for decoding.

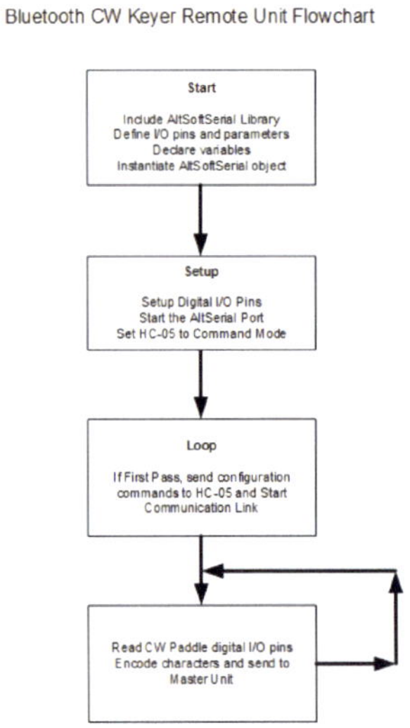

Figure 14.14: The Bluetooth CW Keyer Remote Unit Flowchart.

The sketch for the Remote unit starts out very similar to the Master unit. We'll include the AltSoftSerial library to communicate with the HC-05 module, instantiate the AltSoftSerial object, define the I/O pins and other parameters we'll need, then declare the variables. Again, we need to declare only a handful of variables for this sketch, six in total versus just the four we needed for the Master unit's sketch.

```
#include <AltSoftSerial.h>

AltSoftSerial altSerial;

#define serial_baud 38400
#define bt_baud 38400
#define BT_PWR 3                // Define the Bluetooth module Power Control Pin
#define BT_EN 4                 // Define the Bluetooth module Enable Pin
#define dit_pin 6               // Define the Dit Relay drive pin
#define dah_pin 7               // Define the Dah Relay drive pin
#define command_delay 1000      // Delay between the Bluetooth commands
#define debounce_delay 20 // CW Paddle debounce delay
```

```
bool run_once = true;
String rx_line;
bool receiving, dit_switch, dah_switch;
bool dit_flag = false;
bool dah_flag = false;
char c;
```

In the setup() loop, we set the pin modes for the CW keyer paddle contacts and the Bluetooth module's EN and power control pins. We'll then start the AltSerial serial port at 38,400 baud.

```
void setup()
{
  pinMode(dit_pin, INPUT_PULLUP); // Setup the Dit paddle pin for inout with
pullup resistor
  pinMode(dah_pin, INPUT_PULLUP); // Setup the Dah paddle pin for inout with
pullup resistor
  pinMode(BT_PWR, OUTPUT);        // Setup the Bluetooth module Power Control pin
for output
  pinMode(BT_EN, OUTPUT);         // Setup the Bluetooth Enable pin for output


  altSerial.begin(bt_baud);       // Start the Bluetooth module serial connection
  delay(1000);
```

Finally, we'll set the HC-05 EN pin HIGH and turn on the power to the HC-05 to put it into Command mode.

```
// Turn on Bluetooth PWR and EN pins
  digitalWrite(BT_EN, HIGH);
  delay(command_delay);

  digitalWrite(BT_PWR, HIGH);
```

In the main sketch loop(), on the first pass through, we'll configure the HC-05 module. First, we'll set the Name of the HC-05 module to HC-05, then we'll set the Role to the Slave (remote) role, the Connect Mode to Any, and finally, set the port speed for the connection to the Arduino's Software Serial port to 38,400 baud. Then we'll set the HC-05's EN pin to LOW and send the Reset command to the HC-05 to switch it into Data mode. Finally, we send an ASCII "0" character to the Master unit to ensure that the keying relays in the Master unit are turned off.

When the Bluetooth link is established, the STATE pin on the HC-05 module goes HIGH and turns on the green LED on the circuit board.

```
void loop()
{
  if (run_once)
  {
    delay(command_delay);

    altSerial.println("AT+NAME=HC-05"); // Send the Set Name command
    delay(command_delay);         // Read any responses from the Bluetooth module
    read_bt();                    // Read any responses from the Bluetooth module

    altSerial.println("AT+ROLE=0");    // Send the Slave Role command
    delay(command_delay);
    read_bt();                    // Read any responses from the Bluetooth module

    altSerial.println("AT+CMODE=1");  // Send the Connect to Any command
    delay(command_delay);
    read_bt();                   // Read any responses from the Bluetooth module

    altSerial.println("AT+UART=38400,0,0"); // Send the Baud Rate command
    delay(command_delay);
    read_bt();                   // Read any responses from the Bluetooth module

    delay(command_delay);
    digitalWrite(BT_EN, LOW);    // Turn off Enable pin
    delay(command_delay);
    read_bt();                   // Read any responses from the Bluetooth module

    // Start communication with Master
    altSerial.println("AT+RESET");    // Send the Bluetooth module Reset command
    delay(command_delay);
    read_bt();                  // Read any responses from the Bluetooth module

    run_once = false;

    //  Send an ASCII "0" (zero) to turn off relays on Master
    altSerial.println("0");
  }
```

We did all that setup just to get to this small portion of the sketch. This is the part of the sketch that will sense the Arduino's digital I/O pins that are attached to the CW keyer paddles. Note that this sketch does not actually do a logical AND to the CW keyer contact closure information. Instead, it just sends an ASCII "1" if the "dit" keyer contact is closed and an ASCII "2" if the "Dah" contact is closed. The logical ANDing process occurs on the Master unit based on the state of the keying relays and the incoming ASCII character commands. If both keyer contacts are closed, the Remote unit will send an ASCII "1" followed by an ASCII "2" to indicate both contacts are closed to allow the CW keyer in the transmitter to retain its usual iambic keyer functionality. Note that the ASCII character that indicates a key closure is only sent once and will not be sent a second time unless the keyer contact is released.

```
// Read and send the switch closure information

  dit_switch = !(digitalRead(dit_pin)); // Read the Dit Paddle Input
  dah_switch = !(digitalRead(dah_pin)); // Read the Dah Paddle Input

  if (dit_switch && !dit_flag)      // Handle a Dit paddle press
  {
    altSerial.print("1");           // Send an ASCII "1" to the Master
    dit_flag = true;                // Only send one character until it's released
    delay(debounce_delay);
  }

  if (dah_switch && !dah_flag)     // Handle a Dah paddle press
  {
    altSerial.print("2");          // Send an ASCII "2" to the Master
    dah_flag = true;               // Only send one character until it's released
    delay(debounce_delay);
  }

// Send an ASCII "0" after either paddle was pressed and then released
  if ((!dit_switch && dit_flag) || (!dah_switch && dah_flag))
  {
    altSerial.print("0");
    dit_flag = false;
    dah_flag = false;
    delay(debounce_delay);
  }
}
```

In the last part of the sketch for the Remote unit, we have the read_bt() function. This function will read the incoming serial data from the HC-05 module and is used to send data to the Arduino's Serial Monitor when debugging is enabled.

```
void read_bt()  // Function to read the Bluetooth module responses
{
  rx_line = "";
  do
  {
    if (altSerial.available())
    {
      receiving = true;
      c = altSerial.read(); // Get the incoming character
```

```
      rx_line = rx_line + c;
    }
  } while (altSerial.available());
}
```

Putting it to use

Now that we actually have a working prototype, it comes time to answer those lingering design questions from the design concept phase of this project. To review, those questions were "Can the Bluetooth wireless serial link running at 38,400 baud send the keying information fast enough so that a fast CW operator won't overrun the link speed?" Some test code was inserted into the working prototype sketches and verified that you could indeed send CW keying information at better than 60 words a minute without any issues. I think that pretty much answers the speed question.

Next up was the latency concerns. How badly would the overall latency from paddle key contact closure to actually keying the transceiver impact the whole process? I was actually very surprised when I hooked up the scope to look at the delay time from when the keyer paddle contact was closed and the keying relay on the Master unit was turned on as shown in Figure 14.15. There's a mere end-to-end delay of 20ms across the entire link. I think it's fairly safe to say that latency won't be an issue, and even the fastest CW operator won't notice the latency across the Bluetooth link. Amazingly, this simple design worked perfectly and no major changes were needed from the original concept to the finished project. That of itself is amazing, usually there's always something I overlook that has to be corrected. This project worked as planned right from the start.

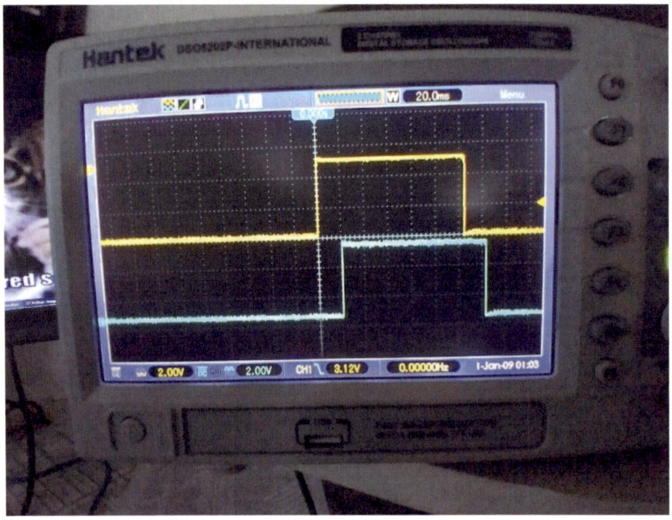

Figure 14.5: Oscilloscope view showing the 20 ms latency between keyer contact closure and the actual keying of the transmitter.

Enhancement Ideas

The original version of this project used at ARRL's Headquarters ham shack was built using an Arduino UNO and prototyping shields for the Master and Remote units. For this book, the project was downsized to use an Arduino Nano. Why stop there? This would be an ideal project for an ESP32 board and have everything fit on a circuit board just big enough to hold the microcontroller and the relays for the Master unit. You can make the Remote unit even smaller and fit it all on a circuit board barely larger than the Arduino/microcontroller proper.

While this project is for a wireless CW keyer, there's no reason not to turn this entire project into a Bluetooth-based wireless telemetry and control system, with bidirectional telemetry and control traffic over the Bluetooth link. There's all kinds of possibilities using this platform as a starting point.

The sketches and libraries for this project are available online at www.kw5gp.com/Elektor.

Chapter 15 • Project 9 — Station Power Monitor

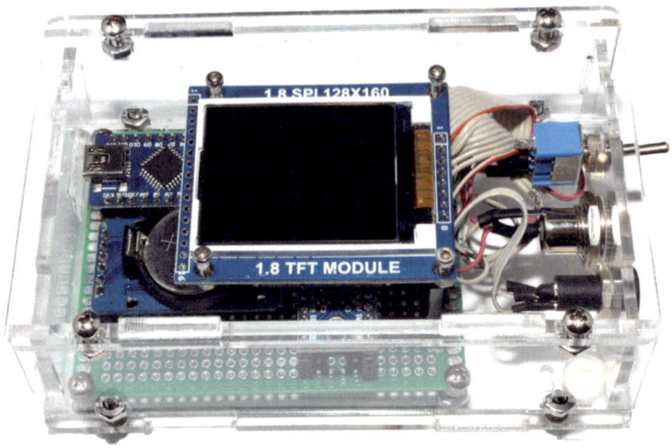

Figure 15.1: The finished Station Power Monitor.

Figure 15.2: The Station Power Monitor display.

During the design and testing phase of this project, I experimented with two methods of measuring the DC voltage and current that I felt worked equally well. In the end, I chose to go with the Hall-Effect version, since it could be used in either a High-Side or Low-Side configuration without no effect to the circuit being measured. (See the sidebar for the difference between High-side and Low-side current measurement).

High-Side versus Low-Side Current Measurement

There are two primary methods of measuring current in a circuit. These are called High-side and Low-Side measurement, based on the placement of the measuring device in the circuit. With High-side measurement, the current measurement takes place at a point between the power supply and the load, while Low-Side measurement takes place between the circuit ground and the power supply ground.

Often, a very low value precision resistor, also known as a shunt resistor, is used as a measurement device. The voltage measured across the shunt resistor will allow you to use Ohm's Law to calculate the current flowing in the circuit. In a typical 50A shunt resistor, the resistance is .001 Ω, with the measurement voltage across the shunt ranging from 0 to 75 mV, depending on the current flow in the circuit. Often, an instrumentation amplifier is used to amplify this minute voltage into a more usable value of 0 to 5 volts for use with the Arduino.

However, there are some concerns that have to be considered when using shunt resistors. When used in a high-side configuration, the instrumentation amplifier must be able to handle the voltage supplied by the power source. As an example, if we place a shunt in the High-side configuration on a 12-volt power supply circuit to a load, one side of a 50-A shunt resistor will be at 12 volts, and the other side will be at 11.925 volts, assuming a load of 50 amps. Even though the differential is only 75 mV, we still have to deal with the 12-volt potential being applied to the input of the instrumentation amplifier, even if the amplifier is used in differential mode. This is easily solved by using voltage dividers on the inputs to the instrumentation amplifier, but it is something to be considered when using High-side measurement.

The issue of high voltage on the input of the instrumentation amplifier is not a concern in Low-side measurement, with the shunt resistor being placed between the equipment ground and the power supply ground. This placement, however, creates its own set of headaches. With the shunt resistor in the Low-side measurement configuration, all of the equipment will be "grounded" at the higher voltage potential side of the shunt resistor, in effect created a condition known as a "floating" ground. Since the voltage across the shunt resistor varies with the current flowing through it, Ohm's Law tells us that the equipment ground can vary from 0 to 75 mV above the power supply ground when using a shunt resistor between the equipment ground and the power supply ground. While this situation, in and above itself, is not really that big of a thing, you also have to add in resistance of the few feet of power wiring for the ground side of your equipment. As an example, #12 AWG wire has a resistance of .0016 Ω per foot (30 cms), which is more than that of the shunt resistor itself. Some equipment is very sensitive to floating ground conditions, along with the associated "grounding loops" that can often occur in a floating ground circuit configuration. Also, this presents issues when used in a chassis-ground system, such as a car, where everything is grounded, and you'd have to isolate all of your equipment from the chassis ground of the car. Because of its placement in the circuit, Low-side measurement also can't detect short circuits to the chassis ground, as such a short circuit would bypass the shunt measurement circuit entirely.

So, while Low-side measurement may seem to some have advantages over High-side measurement as far as dealing with high voltages, at the end of the day, in most cases, High-side measurement is generally considered to be the better choice of the two, when possible.

Of course, you can avoid all of the downsides of either method by using a non-invasive measuring device such as a Hall-Effect sensor, which is why the Hall-Effect sensor version of the Station Power Monitor is presented here.

This design uses a 30-amp Hall-Effect Current Sensor module to measure the current in the circuit. While the sensor module is actually connected to the circuit, the only part of the module inserted into the actual circuit is the circuit board trace and the ACS712 Hall-Effect sensor chip input and output pins that act as just a short piece of wire as far as the circuit is concerned. This "wire" is connected to the input and output pins of the ACS712 Hall-Effect sensor chip, which uses the magnetic field induced between the pins to measure the current while remaining isolated from the circuit under test. In this manner, we don't need to be concerned about the actual circuit voltage, as the chip itself offers a minimum of 2.1 kV of isolation.

While I like the circuit isolation provided by using the Hall-Effect sensor module, the sensor module is available in only 3 ranges, 5, 20, and 30 amps. To read higher currents, we'd most likely need to use an external shunt resistor and read the voltage across the shunt resistor, using Ohm's Law to determine the current flow through the shunt resistor. Of course, this method requires the part of the project that measures that voltage to be a part of the circuit under test and could run into the voltage and grounding issues as mentioned in the sidebar discussion. To keep things simple, we'll use the 30-A version of the ACS712 Hall-Effect module in this project and save the current shunt measuring for another day.

Design Goals

My concept of a Station Monitor goes back to the days of the Heathkit SB-630 Station Console. While the SB-630 had an SWR meter, and a phone patch, in addition to a digital clock, something similar is what I had in mind for this project. I wanted to build a small station accessory box that would use a color TFT display to show voltage, current, date, and time. Since the majority of rigs today use 12-volts DC for power, the voltage and current readings would be taken from the station's 12-volt power supply. A DS3231 real-time clock calendar (RTC) module would be used to keep track of the date and time. Interfaced using the I^2C bus, the DS3231 module is an extremely accurate RTC, with a built-in temperature-compensated crystal oscillator (TCXO), and it uses a small coin-cell battery to maintain timekeeping when the power is turned off. The DS3231 also has a built-in temperature sensor we can use to display ambient temperature on our station monitor display.

Since we're using a color TFT display, we'll change the color of the text on the display when there is an over/under voltage, or high current condition. To personalize the station monitor, we'll add our callsign to the display. The entire project will fit inside a Solarbotics MegaSAFE enclosure, so we'll use an Arduino Nano mounted on the usual 60 mm by 80 mm piece of copper clad perfboard, which we'll mount to the base of the enclosure. For the display, we'll use an ST7735-type 160 x 128 pixel 1.8 inch color TFT display.

Designing the project

A Hall-Effect sensor uses the magnetic field created by current flowing through a wire to determine the current flowing in a circuit. The intensity of the magnetic field is proportional to the current flowing through the wire. A Hall-Effect sensor will produce a similar output based on the intensity of the magnetic field generated by the wire, without having to actually be inserted in the circuit. This type of current sensing is known as non-invasive because the actual sensing portion of the device is not in-line with the circuit that that is being measured. This type of design provides complete isolation between the circuit being tested and the measuring device.

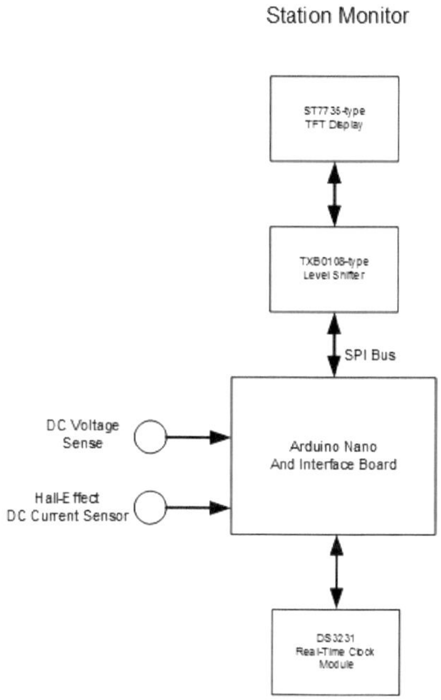

Figure 15.3: The Station Power Monitor block diagram.

Figure 15.3 shows the Block Diagram for the Station Power Monitor project. We'll use an ACS712 30A Hall-Effect current sensor module, a DS3231 RTC module (Figure 15.4) for our date, time, and temperature, and the usual ST7735-type TFT display. The 1.8-inch display was chosen because it fits nicely inside the Solarbotics enclosure and is mounted to the inside of the clear enclosure lid using 2 mm hardware.

This project uses both the SPI and I^{2}C buses to communicate with the display and RTC module, respectively. The Hall Effect sensor shown in Figure 15.5 uses an Allegro Microsystems ACS712 IC to convert the magnetic field intensity generated by the current flow in the circuit under test into an output voltage. With the 30-A version of the module, the sensor will generate 66mV per Amp on the output pin. The ACS712 module I used is capable of sensing up to 30 A of both positive and negative current, so the output voltage of the module is centered at VCC/2, or 2.5 V. This means that we can feed the output of the Hall-Effect module directly into an analog pin on the Arduino without the need for voltage dividers or an instrumentation amplifier circuit. It is important to note that this form of measurement cannot measure AC current. For that, you'll need a different type of sensor, often a split-core transformer to read the AC current. While the Hall-Effect module is a non-intrusive current sensor, the power supply and load connections for the circuit to be tested are made using screw terminals on the module. This is to allow the current to connect to the ACS712's input pins and flow at a precise distance from the internal Hall-Effect sensor so that it can accurately sense the magnetic field from the current flow. For convenience, the module will connect to the main unit using a length of multiconductor cable to supply the +12 V input voltage, the sensor module voltage output, and the power supply ground connection. Powerpole pigtails were used to allow easy insertion of the sensor assembly between the power supply and the equipment load. The assembly I used is shown in Figure 15.6. Note that the sensor module is usually insulated with a large piece of heat-shrink tubing that was removed to show the sensor placement in the circuit. Four Powerpole connectors were used to create a connector block for the Hall-Effect module and input power connections. The Hall-Effect sensor assembly was placed at the end of this connecting cable to allow you to place the sensor module in-line with your station's power supply, while having the rest of the unit on top of your desk where you can see the display.

Figure 15.4: The DS3231 Real-Time Clock module.

Figure 15.5: The ACS712 Hall-effect Current Sensor.

Figure 15.6: The cable assembly used to insert into my shack's DC power system.

Project Flowchart

The next step after the basic hardware design is to draw the flowchart (Figure 15.7) for what we want our station power monitor to do. Now that we have an idea of the hardware pieces in the project, we can map out what we want to do with those pieces. We'll also need to include and initialize all the libraries we'll be using in this project.

The Adafruit GFX library is the core graphics library that supports many of the color TFT displays, containing high-level functions that allow you to draw lines, circles, rectangles and other objects, and text using a simple set of commands that is common to color TFT displays. The device-specific extension to the GFX library for this project is the Adafruit ST7735 library, which supports the ST7735 graphics controller used by the 1.8 inch TFT. Since the TFT display uses the SPI bus, we may also need to include the SPI library, however, more recent versions of the Adafruit libraries automatically include the SPI library for you. It doesn't hurt to include the SPI library, as duplicate include statements are ignored by the compiler. The last library we'll need is a library for the DS3231 RTC. I chose the DS3231 library from Rinky-Dink Electronics for its simplicity and ease of use.

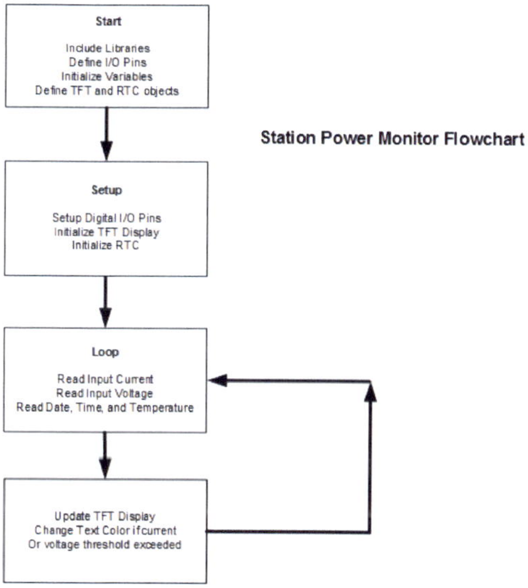

Figure 15.7: The Station Monitor flowchart.

The flowchart for this project is relatively straightforward. After initialization and setup, the sketch will constantly read the voltage and current of the circuit, update the date, time and temperature on the display every 10 seconds. Any voltage or current out of limit conditions will be indicated by changing the text color of the display from green to yellow or red depending on the alarm condition.

There are at least three different versions of the 1.8 inch TFT display, known as the Black, Green, and Red Tab versions. There are internal differences between these various versions and you must initialize the TFT with the correct Tab type. On the front of the TFT display, near the ribbon connector that connects the display to the display module board you will see a small plastic tab. The color of this tab is used to determine the Tab type of ST7735-type display module. If you don't use the correct Tab type when initializing the display, you may encounter such issues as poor display quality, or incorrect colors. Initializing with the incorrect Tab type will not damage the display module, but it can definitely cause display issues.

Schematic

Figure 15.8 shows the schematic and Figure 15.9 shows the Parts List for the completed Station Power Monitor project. Once the design was complete and tested on my bread-board, I constructed the project on a 60 mm by 80 mm piece of copper-clad perfboard as shown in Figure 15.10. The board was then mounted to the base of the Solarbotics Enclo-sure.

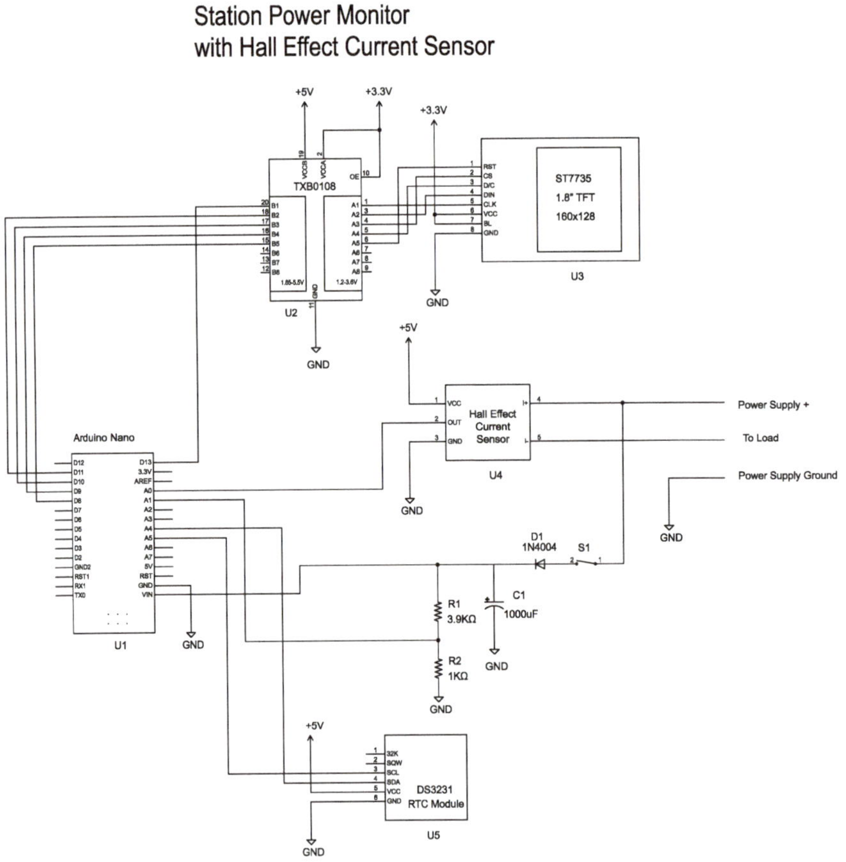

Figure 15.8: The Station Monitor schematic.

Parts List
Station Monitor

Part	Value
C1	1000 µF, 25 V capacitor
D1	1N4004 diode
S1	SPST switch
R1	3.9 kΩ resistor 1/8W
R2	1 kΩ resistor 1/8W
U1	Arduino Nano
U2	8-channel Level Shifter Module (TXB0108 type)
U3	ST7735-type 1.8" color TFT display
U4	1A169 Hall-effect current sensor
U5	DS3231 RTC/Temperature module
Enclosure	Solarbotics Mega S.A.F.E.

Figure 15.9 The Station Monitor parts list.

Figure 15.10: The Station Monitor circuit board.

The Sketch

The Station Power Monitor sketch starts with including the Adafruit ST7735 and DS3231 libraries. The SPI library is included as well, but more recent versions of the Adafruit libraries include the SPI library automatically and this step may not be necessary. It is included here just to be sure the library is included.

```
#include <Adafruit_GFX.h>     // Core graphics library
#include <Adafruit_ST7735.h> // Hardware-specific library
#include <SPI.h>
#include <DS3231.h>
```

Next, we'll define the I/O pins for the TFT, the input pin for the Hall-Effect sensor, and the voltage sense pin for the voltage divider on the power supply input. This project will use the standard hardware SPI pins on the Nano to provide the best speed on display updates.

```
// For the breakout, you can use any 2 or 3 pins
// These pins will also work for the 1.8" TFT shield
#define TFT_CS     10
#define TFT_DC     9
#define TFT_RST    8  // you can also connect this to the Arduino reset
// in which case, set this #define pin to -1

Adafruit_ST7735 tft = Adafruit_ST7735(TFT_CS,  TFT_DC, TFT_RST);
#define voltage_pin A1
#define current_pin A0
```

Next, we'll initialize the DS3231 I^2C Real-time Clock module. You may notice that the SDA and SCL pins are not defined in the sketch. The hardware libraries for the Arduino Nano (and most other Arduinos as well) have already defined the correct SDA and SCL pins, so you can just use the SDA and SCL nomenclature in the initialization statement.

```
// Init the DS3231
DS3231 rtc(SDA, SCL);
```

The 30-A Hall-Effect Sensor module can sense both positive and negative 30 A. This means that at zero current, the output voltage of the module is VCC/2, or 2.5 V. Our sketch will need to take this offset into account and save both the offset A/D value and the maximum current output value for use later in the sketch.

```
float current_cal_0 = 3.0;
int current_cal_max = 1009;
```

The last part of the sketch initialization is where we define the variables to be used in the sketch. We'll create the variables used to contain the voltage and current readings, along with calibration settings that will be used to properly calculate the voltage and current. The threshold values are used to change the color of the display to indicate any warning conditions. The high current warning level is set for 25 A, the low voltage warning level is set for 11.9 V, and the high voltage warning level is set for 14.0 V. You can adjust these values as needed for your personal preferences.

```
float voltage, current;
float volt_drop = 0.776;
int volt_cal = 2480;

float current_high = 25;
float voltage_low = 11.9;
float voltage_high = 14.0;
```

```
float temperature;
String rtcdate, rtctime, rtcday, rtcmonth, rtcyear ;
```

Now it's time to do the setup() loop. The first thing we do is initialize the TFT color display. Remember that you'll have to initialize your TFT display with the correct Tab type, in my case, BLACKTAB.

```
// Use this initializer if you're using a 1.8" TFT
  tft.initR(INITR_BLACKTAB);   // initialize a ST7735S chip, black tab
```

Next, we'll need to initialize the DS3231 RTC module. You'll notice that after we start the RTC, the statements to set the Day of Week, Time, and Date are commented out. You will need to uncomment these statements, enter the correct information and upload the sketch to set the date and time. After the date and time are set, you'll need to comment these lines out and upload the sketch again, otherwise the RTC will revert to the date and time settings in these statements every time you reset or power on the Arduino.

```
// Initialize the rtc object
  rtc.begin();

  // The following lines can be uncommented to set the time
  // rtc.setDOW(FRIDAY);          // Set Day-of-Week to SUNDAY
  // rtc.setTime(11,12,0);        // Set the time - format is h:m:s (24hr format)
  // rtc.setDate(4,8,2023);       // Set the date - format is dd.mm.yyyy
```

Now we'll start setting up the TFT display, clearing it, and setting the background color to blue. Note that when you "clear" a TFT display, you actually just fill the screen with the background color. Anything you write to the display afterward keeps overwriting the exiting displayed information until you fill all, or part, of the display with the background color. We'll also need to rotate the display by 90°, since the TFT is actually mounted on its side in our project. We'll set the desired font size, set the text color to green, and display the startup screen for 5 seconds.

```
  tft.fillScreen(ST7735_BLUE); // Clear the display
  tft.setRotation(1);          // Set the screen rotation
  tft.setTextWrap(false);
  tft.setTextSize(2);
  tft.setTextColor(ST7735_GREEN);
  tft.setCursor(50, 30);
  tft.print("KW5GP");
  tft.setCursor(37, 60);
  tft.print("Station");
  tft.setCursor(37, 80);
  tft.print("Monitor");
  delay (5000);
  tft.fillScreen(ST7735_BLUE);
```

In the main loop(), the sketch will clear the display, set the text color to green, and display the callsign at the top of the TFT display. Next, we'll execute a simple do loop, where the sketch will execute this loop endlessly. Every 10 seconds, the sketch will repeat this loop, and read the sensor and RTC values and update the display. As part of the display update, we'll check for any warning threshold conditions, and set the display text color accordingly. A high current condition will be displayed in red, a low voltage condition is displayed in yellow, and a high voltage condition is displayed in red, otherwise the value is displayed in green. The DS3231 RTC module also has a built-in temperature sensor. We'll read that and show the ambient temperature on the display as well.

```
void loop()
{
  tft.setTextSize(3);
  tft.setTextColor(ST7735_GREEN);
  tft.fillScreen(ST7735_BLUE);
  tft.setCursor(40, 5);
  tft.print("KW5GP");

  do
  {
    tft.setTextSize(3);
    tft.setCursor(30, 40);

    readCurrent();
    readVoltage();
    tft.setTextColor(ST7735_GREEN, ST7735_BLUE);
    if (current >= current_high)
    {
      tft.setTextColor(ST7735_RED, ST7735_BLUE);
    }
    tft.print(current, 2);
    tft.print("A  ");
    tft.setCursor(30, 70);
    tft.setTextColor(ST7735_GREEN, ST7735_BLUE);
    if (voltage <= voltage_low)
    {
      tft.setTextColor(ST7735_YELLOW, ST7735_BLUE);
    }
    else
    {
      if (voltage >= voltage_high)
      {
        tft.setTextColor(ST7735_RED, ST7735_BLUE);
      }
    }
```

```
    tft.print(voltage, 2);
    tft.print("V  ");
    tft.setTextColor(ST7735_GREEN, ST7735_BLUE);
    tft.setTextSize(2);
    tft.setCursor(55, 100);
    temperature = rtc.getTemp();
    temperature = (temperature * 1.8) + 32;
    tft.print(temperature, 1);
    tft.print(" F ");
    tft.setTextSize(1);
    tft.setCursor(0, 120);
    rtcdate = rtc.getDateStr();
    rtcmonth = rtcdate.substring(3, 5);
    rtcday = rtcdate.substring(0, 2);
    rtcyear = rtcdate.substring(6, 10);
    rtctime = rtc.getTimeStr();
    rtctime = rtctime.substring(0, 5);
    tft.setCursor(0, 120);
    tft.print(" ");
    tft.print(rtc.getDOWStr());
    tft.print(" ");
    tft.print(rtcmonth);
    tft.print("/");
    tft.print(rtcday);
    tft.print("/");
    tft.print(rtcyear);
    tft.print(" ");
    tft.print(rtctime);
    delay(10000);
  } while (true);
```

The Station Power Monitor has two functions. The readVoltage() function is used to read the input power supply voltage from the power supply voltage divider. The A/D value of the analog input pin is converted into the proper voltage for display on the TFT.

```
void readVoltage()
{
  float count = analogRead(voltage_pin);
  voltage = map(count, 0, 1023, 0, volt_cal);
  voltage = (voltage / 100) + volt_drop; // Map the Voltage A/D value to voltage
}
```

The readCurrent() function will read the voltage on the analog input pin for the Hall-Effect sensor output. Since the Hall-Effect current sensor can read both positive and negative current values, we'll need to apply the offset calibration value to determine the zero current value. Since our project is designed to only display positive current values, we will only deal with positive current values higher than the zero current calibration point. We'll then use a map statement to convert the A/D value to the correct current value, in ampères, to be displayed.

```
void readCurrent()
{
  float count = analogRead(current_pin);  // read the Current value
  count = count + current_cal_0; // set the zero center point at approx 2.5V
  current = map(count, 0, current_cal_max, -3000, 3000); // convert it to Amps
XX.X
  current = current / 100;
}
```

Enhancement Ideas

This project lends itself to a number of modifications and enhancements. Setting the date and time on the DS3231 requires setting the date and time in the sketch, uploading it, commenting out the data and time setting statements, and uploading the sketch again. Updating the sketch to allow entering the current date and time, if desired, would be very helpful. Using a GPS module instead of the DS3231 RTC module would eliminate the need for setting the date and time altogether, as well as providing very accurate time. With a GPS instead of the RTC, you could even add your current grid square to the display.

If you still wanted the ambient temperature display, you could add a DS18B20 temperature sensor in addition to the GPS module. You could also add equipment protection to the project by adding a shut-off relay in line with the current sensor and the load side to turn off the power to your equipment in the event of an over or undervoltage condition. And finally, you could add speech capability using a text-to-speech module.

The sketch and libraries for this project are available online at: www.kw5gp.com/Elektor.

Chapter 16 • Project 10 — AC Current Monitor

Figure 16.1: The finished AC Current Monitor.

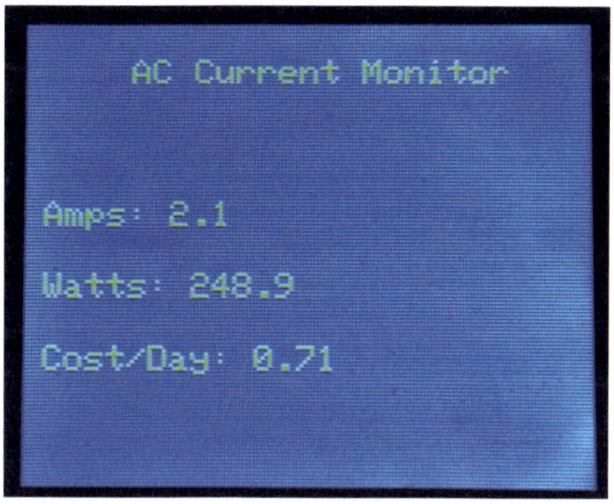

Figure 16.2: The AC Current Monitor display.

On a previous job as a network engineer and technology consultant, I was tasked with preparing a seminar on cost savings. Using P3 International's *Kill-A-Watt^{tm} EZ Power Meter* (Figure 16.3), I was able to calculate the average power consumption and cost of running a PC workstation and monitor on a kilowatt/hour (kW/h) basis. While individually, you don't think a whole lot about how much power your PC and monitor consume, when you scale that up to the estimated 150,000 workstations in the network at that time, any cost savings would add up in a hurry. At that time, workstations and monitors were being left on 24 hours a day, primarily because of the myth (busted by MythBusters, in fact) that you used more electricity to turn a device on and off than you did if you just left it on, and also a belief that leaving the workstation and monitor on was less stressful than turning it on at the start of the day, and off again at the end of the day.

Figure 16.3: The Kill A Watt EZ AC power monitor.

Needless to say, the resulting projected cost savings was rather impressive, and since then, a number of companies now deal exclusively in cost savings measures and technology to turn such devices off after hours. There are actually far easier ways to do that, but that's a topic for another day. Naturally, when it came time to think up the projects for this book, that AC Kilowatt/hour meter came to mind. The *Kill-A-Watt*^{tm} EZ design required that you insert it between the wall outlet and the device to be measured. With the relatively low cost of a non-invasive AC current sensor, such as the Dechang Electronics YHDC SCT013 series of split-core AC current transformers (Figure 16.4), I could build a more versatile, non-invasive version of the *Kill-A-Watt*^{tm} EZ.

Figure 16.4: A non-invasive current sensor.

Design Goals

The design goals for this project are relatively simple. We want to be able to non-invasively read and display the AC current, Watt usage, and an estimated daily cost in real-time. Since I wanted a portable, rugged, unit, I chose to use an Arduino Nano mounted on a 60 mm by 80 mm piece of prototyping board attached to the bottom panel of a Solarbotics MegaSAFE enclosure with a 1.8-inch ST7735-type color TFT Display module mounted to the inside top cover of the enclosure using 2-mm hardware. The AC Current Transformer will connect to the Arduino through an Analog Devices AD623 Instrumentation Amplifier IC circuit. It is important to note that a split-core transformer design can only measure AC current as transformers, by design, block DC voltage and current.

The YHDC SCT013 series of split-core AC Current transformers are precision transformers that have a small latch on the side, allowing you to open the transformer, insert one leg of the AC circuit, and close the transformer around the wire. When measuring AC current using a non-invasive sensor such as this, you can't put both the hot and neutral wires inside the sensor, as their magnetic fields would cancel each other out. The YHDC SCT-013 15-amp current sensor used in this project is a 1 to 1800 turn transformer that will convert the magnetic field created by the AC current into an AC voltage varying from 0 to 1 Volt RMS based on the amount of current. The SCT013 series has a range of current options, from 5 A to 100 A. I chose the 15-A version for this project mainly because 15 amps is the rating for most wall power outlets.

Be aware that there are two versions of the SCT13 series of current transformers. The SCT13-000 outputs current instead of voltage. You will want to use the SCT13-XXX series where XXX is the current range you wish to read. These sensors output a voltage from 0 to 1 volt based on the current flowing in the circuit under test. The voltage version of these sensors also has the "burden" resistor built-in. For the current output version, a burden resistor is required, and is placed across the output of the transformer in order to generate the output voltage from the sensor.

Since the current transformer outputs an AC signal from 0 to 1 volt, we will need to apply a DC offset to the waveform to ensure that we don't have the instrumentation amplifier trying to deal with the negative portion of the sine wave. Fortunately, the AD623 has a reference voltage pin that we can use to apply a DC offset to the input signal. This will allow us to amplify the entire sine wave coming from the sensor. On the output of the AD623, we will add a simple rectifier circuit to convert the AC signal into a DC voltage that we can read using the Arduino's A/D pin. Theoretically, we could sample the incoming sine wave directly on the Arduino's analog input pin, but then we would have to do the AC voltage conversion in software. It's far easier and less software intensive to rectify the signal in hardware ahead of time.

The Design

The Block Diagram for the AC Current Monitor is shown in Figure 15.5. The simplicity of the design is a testament to the versatility of the Arduino for basic sensing and control tasks. To interface the AC current sensor to the Arduino, we'll use an AD623 instrumentation amplifier circuit, and the power switch will be mounted on the side of the enclosure itself.

Since the AC Current sensor has a stereo mini-plug on the end of its cable, we'll mount a stereo mini-jack on the side of the enclosure to allow using different current ranges of the sensor simply by plugging in a different sensor and modifying the sketch to support the new current range.

The rectifier circuit for the sensor input is a simple half-wave rectifier. The 10uF capacitor on the output of the rectifier will help filter out most of the remaining AC ripple on the input to the Arduino's analog input pin. The 1MΩ resistor in the rectifier circuit is to help "bleed" down the filter capacitor charge as the AC current changes. This allows a faster DC voltage response as the current varies. The input impedance of the Arduino's A/D converter as listed on the ATMega328 datasheet as 100 MΩ, so it takes a bit of time for the Arduino A/D's internal sample and hold capacitor to settle to a new voltage as the current varies. We use the 1-MΩ resistor to help the A/D to respond a little bit faster to changes in input voltage without affecting the actual voltage level itself.

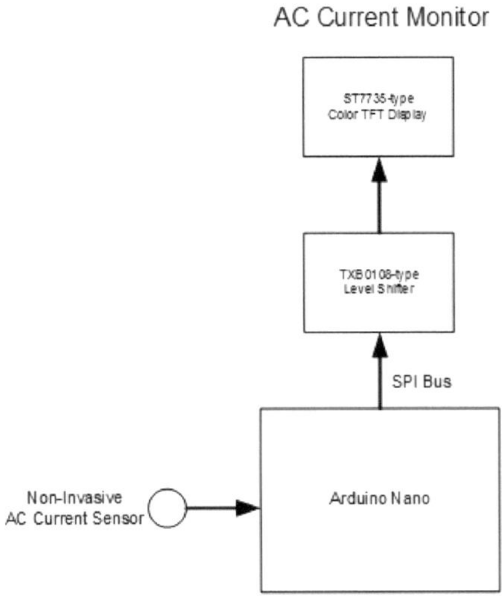

Figure 16.5: The AC Current Monitor block diagram.

As with the block diagram, the Flowchart for the AC Current Monitor (Figure 16.6) is also simple and straightforward. After we initialize and set up the Arduino I/O pins and the TFT display, we'll read the input current from the current sensor, calculate the Wattage and the estimated cost of the KW/h usage. We'll then delay 5 seconds before repeating the cycle.

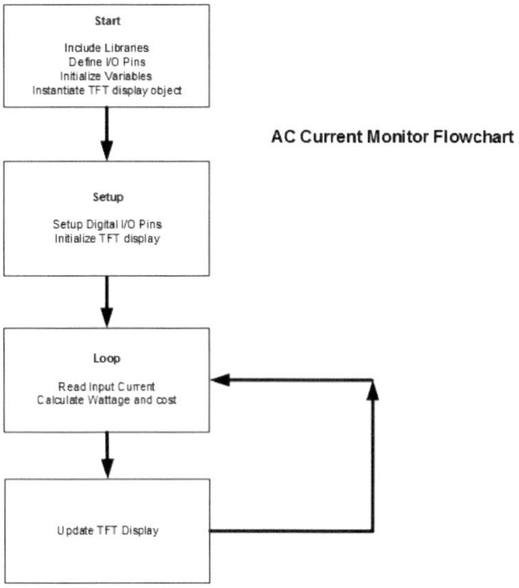

Figure 16.6: The AC Current Monitor Flowchart.

The schematic diagram for the AC Current Monitor is shown in Figure 16.7, and the Parts List in Figure 16.8. As you can see, there is not a lot of external wiring needed to build this project, and it can easily be built in an afternoon.

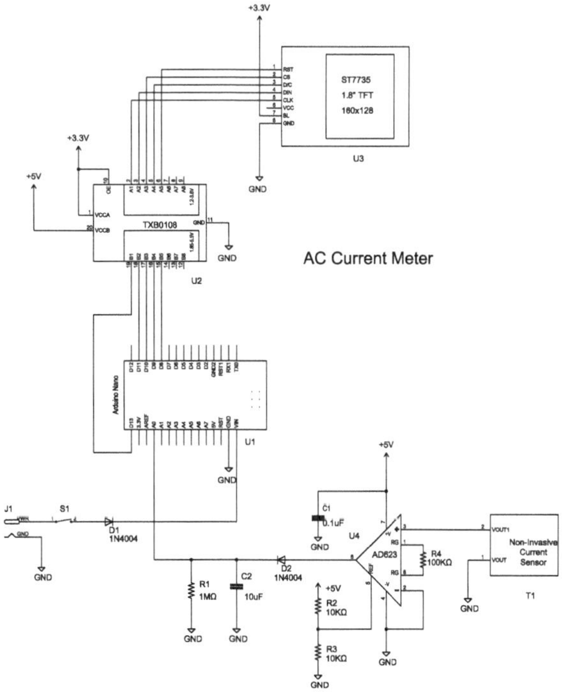

Figure 16.7: The AC Current Monitor schematic.

Parts list
AC Current Monitor

Part	**Value**
C1	0.1 µF 16 V capacitor
C2	10 µF 16 V capacitor
D1, D2	1N4004 diode
R1	1 MΩ resistor, 1/8W
R2, R3	10 kΩ resistor, 1/8W
R4	100 kΩ resistor, 1/8W
S1	SPST Switch
T1	SCT013 non-invasive AC current sensor
U1	Arduino Nano
U2	TXB0108 8-channel level shifter module
U3	ST7735 1.8" color TFT display
Enclosure	Solarbotics Mega S.A.F.E.

Figure 16.8: The AC Current Monitor parts list.

As with as many chips and modules as possible in my projects, the AD623 chip is mounted in a socket, just in case. DuPont-style 0.1-inch (2.54-mm) socket headers are used as sockets for the Arduino Nano and the TXB0108 Level Shifter. DuPont-style 2.54 header pins are used for the connections between the TFT display, current sensor mini phone jack, and power switch as shown in Figure 16.9.

You may notice that I almost always use a polarity protection diode in my Arduino power circuit. Since I use connectors for just about everything in my Arduino projects, including power, it's all too easy to reverse the connections. The diode will help protect the Arduino circuit in case I connect the power backwards.

Figure 16.9: The AC Current Monitor circuit board.

The Sketch

The sketch for the AC Current Monitor is straightforward and relatively simple. In the initialization section, we'll include the Adafruit GFX and ST7735 libraries for the color TFT display. We'll also define the Current sensor pin to be the Arduino's analog input pin A0. We'll also define the zero current point offset and the maximum full scale current A/D values for the sketch. You can adjust these calibration values for your project as needed. Then we'll define the TFT I/O pins and define a 10-ms delay to be used between TFT display commands. This delay is necessary because some of the third-party versions of the ST7735-type display aren't able to update as quickly and often need this brief delay to give the modules a little extra time to process the display commands.

```
#include <Adafruit_GFX.h>      // Core graphics library
#include <Adafruit_ST7735.h>   // Hardware-specific library

#define Current_Pin A0         // Define the Current Sense pin
#define zero_point 420         // Define the Zero Setpoint
#define max_point 509          // Define the High (15A) setpoint

#define TFT_CS     10
#define TFT_DC     9
#define TFT_RST    8

#define tft_delay 10           // set the TFT command delay to 10ms
```

Next, we'll initialize the variables used in the sketch as well as configuring the sketch for the AC voltage of the circuit to be tested. I measured the AC voltage on the wall outlet in my shack to set this value in the sketch. I then used my electric bill to calculate the cost per kilowatt/hour (kWh). In my case I included all the fees and taxes to get a more realistic number. My cost per kW/h is 11.83 cent. More than likely, you'll need to adjust this value for the electricity costs in your area.

```
float watts, amps, cost_day;   // Variables for wattage, amperage, and the
estimated cost per day
int count;                     // The A/D count of the Current Sense pin

// Change these for your location
const float voltage = 119.1;   // The AC line voltage
const float cost_kwh = 0.1183; // Cost per kilowatt/hour
```

In the last part of the initialization area, we'll instantiate (create) the TFT display object.

```
// Instantiate the TFT Display
Adafruit_ST7735 tft = Adafruit_ST7735(TFT_CS,  TFT_DC, TFT_RST);
```

The setup() loop for the AC Current meter is also short and sweet. Since the Arduino's analog input pins default to analog input, we don't even have to set the pin mode. When we issue the analogRead() command, the specified analog input pin will also automatically be placed in analog input mode as part of the analogRead() operation. So, all we have to do in the setup loop is to initialize and clear the TFT display and display a startup screen for 5 seconds.

```
void setup()
{
  // Use this initializer if you're using a 1.8" TFT
  tft.initR(INITR_BLACKTAB);   // initialize a ST7735S chip, black tab
  delay(tft_delay);
  tft.fillScreen(ST7735_BLUE); // Clear the display
  delay(tft_delay);
  tft.setRotation(1); // Set the screen rotation
  delay(tft_delay);
  tft.setTextWrap(false);
  delay(tft_delay);
  tft.setTextSize(2);
  delay(tft_delay);
  tft.setTextColor(ST7735_GREEN);
  delay(tft_delay);
  tft.setCursor(50, 30);
  delay(tft_delay);
  tft.print("KW5GP");
  delay(tft_delay);
  tft.setCursor(20, 60);
  delay(tft_delay);
  tft.print("AC Current");
  delay(tft_delay);
  tft.setCursor(40, 80);
  delay(tft_delay);
  tft.print("Monitor");
  delay(tft_delay);
  delay (5000);
  tft.fillScreen(ST7735_BLUE);
  delay(tft_delay);
  tft.setTextSize(1);
  delay(tft_delay);
  tft.setCursor(30, 10);
  delay(tft_delay);
  tft.print("AC Current Monitor");
}
```

Starting out in the main loop(), we'll read the voltage on the Arduino's analog input pin for the current sensor. Since the zero current point will be approximately VCC/2 or 2.5 volts due to the DC offset voltage we apply to the AD623 instrumentation amp, we'll subtract this DC offset value from the A/D reading. Sometimes, due to noise on the input voltage, the resulting reading may be below zero. We'll have the sketch convert any negative readings to zero since AC current doesn't really have a positive or negative polarity to it, and we're usually only talking about a couple of A/D counts, which will have no noticeable effect on the actual current values.

```
void loop()
{
  count = analogRead(Current_Pin);  // Read the incoming current sense voltage
  count = count - zero_point;       // Subtract the zero point value
  if (count < 0) // Our count should be zero, but if it's negative, make it zero
  {
    count = 0;
  }
```

Next, we'll use a map statement to map the A/D values to a range of 0 to 1500. We use the 1500 value because we want to display the resulting current from 0 to 15 Amps in hundredths of an amp. By using the map statement in this way, the resulting value is trimmed to the correct number of decimal points when we divide by 100.

```
amps = map(count, 0, max_point, 0, 1500); // Map the current sense voltage to
hundredths of an Amp
  amps = amps / 100; // Divide the current by 100 to place the decimal point
```

Now that we have the current reading in amps, we can calculate the wattage and print the current and wattage to the TFT display. We'll then delay for 5 seconds, clear the data area of the display and repeat the main loop().

```
tft.setCursor(5, 50);
  delay(tft_delay);
  tft.print("Amps: ");
  delay(tft_delay);
  tft.print(amps, 1);
  delay(tft_delay);
  tft.setCursor(5, 70);
  delay(tft_delay);
  tft.print("Watts: ");
  delay(tft_delay);
  watts = (amps * voltage); // Calculate and display the wattage
  tft.print(watts, 1);
  delay(tft_delay);
  tft.setCursor(5, 90);
  delay(tft_delay);
```

```
  tft.print("Cost/Day: ");
  delay(tft_delay);
  cost_day = ((watts / 1000) * cost_kwh) * 24; // Calculate and display the
estimated Kilowatt/Hour cost per day
  tft.print(cost_day, 2);
  delay(tft_delay);
  delay(5000);
  clear_data();
```

The clear_data() function is used at the end of the sketch to clear the data area of the TFT display before repeating the sketch loop().

```
void clear_data() //Clears the data area of the display
{
  tft.fillRect(0, 30, 159, 90, ST7735_BLUE); // Clear the CAT data area
  delay(tft_delay);
}
```

To test the AC Current Monitor, I used a modified AC extension cord to allow me to separate out the wires and attach the non-invasive split-core transformer on one wire of the AC circuit. I also verified this current reading with a multimeter inserted into the circuit. Not only did the non-invasive current sensor work well, but it was also reasonably accurate. It was kind of fun to run around the house seeing what all of the various appliances cost to operate.

Enhancement Ideas

As simple as this project appears to be, there are several enhancements that I would consider doing down the road. As the project is designed, the AC voltage used for all the calculations is defined within the program. For more accurate readings, you might want to add the capability to measure the actual AC voltage of the circuit under test and have it dynamically apply that AC voltage reading to the calculations.

Another cool thing you might want to do is to graph the current usage over time. You could either save the data to an SD card and graph it on a workstation using a Processing sketch or you could connect the workstation to the AC Current monitor using the Arduino's USB port and use Processing to graph the current usage and kW/h cost over time. Or even simpler, you can take advantage of the Arduino IDE's new Serial Plotter tool and use the Arduino IDE to display the graph. Finally, you could just simply graph the date to the color TFT display itself.

The sketch and libraries for this project are available at www.kw5gp.com/Elektor.

Chapter 17 • In Conclusion

I hope this book has helped you along the way in your own personal Arduino adventure and that you enjoyed the projects presented here as well. But don't let this be the end of your adventure. The world of the Arduino and microcontrollers is ever-growing and there's always some new board, module, or sensor that will allow you to create amazing and fun projects. If there's anything I can do to help you on your journey, feel free to reach out to me at kw5gp@arrl.net and I'll help as best I can.

Just in the timeframe it took for me to prepare the projects and author this book, many new boards and devices have joined the Arduino and microcontroller family. These new boards have so much more power and speed than the original Arduino UNO and Nano, using 32-bit processors instead of the UNO and Nano's 8-bit processor. With the power and speed of these new boards, creating an Arduino-powered Software-Defined Radio (SDR) is now within reach at a very low cost.

For example, there's the ESP32 (Figure 17.1). The ESP32 has a clock speed of up to 240 MHz and 4 to 16MB of Flash memory, 38 to 77 I/O pins, up to 18 12-bit A/D pins and two D/A pins. The ESP32 also features onboard WiFi and Bluetooth, all for a cost of $3.

Figure 17.1: The ESP32.

Then there's the STM32 "Blue Pill" (Figure 17.2), also for about $3. The Blue Pill has a 72MHz processor with 64KB of Flash and 20KB of SRAM. It has 32 digital I/O pins, 14 analog input pins, and much more.

Figure 17.2: The STM32.

Next, there's the Raspberry Pi Pico (Figure 17.3), with its dual-core 133MHz processor, 2MB of Flash and 256KB of SRAM. The Pi Pico also has 26 digital I/O pins, three 12-bit A/D pins, and includes an on-chip real-time clock and temperature sensor. The Pi Pico even has on-board USB Host support, along with WiFi and Bluetooth. All for about $6.

Figure 17.3: The Raspberry Pi Pico.

The Pi Pico is also leading the way for what I feel is the next wave of microcontroller evolutionary progress. The Pi Pico can support a number of different programming languages with a simple press of the board's BOOTSEL button. At the time of this writing, the Raspberry Pi Pico supports two versions of Python, the Arduino IDE, a C/C++ compiler, and two versions of BASIC.

We'll talk more about Python in a bit, but for now, let's continue with the hardware. Boards are now becoming available with integrated TFT displays. The LILYGO T-PicoC3 (Figure 17.4) has not only an integrated color TFT display on the board, but it has both a Raspberry Pi Pico RP2040 and an ESP32 C3 processor on the same board. All you do is flip the USB-C cable over to switch between the two processors.

Figure 17.4: The LilyGo T-PicoC3.

Arduino is also still creating and releasing new boards, with the recently released Arduino UNO R4 series that includes the UNO R4 WiFi (Figure 17.5) with onboard WiFi and Bluetooth capabilities, the new Nanos with ESP32 and RP2040 processors (Figures 17.6 and 17.7), and even more recently, the GIGA R1 WiFi with a dual-core 480 MHz processor, 18 MB Flash, 1 MB SRAM, 76 I/O pins, and WiFi and Bluetooth support.

Figure 17.5: The Arduino R4 WiFi.

Figure 17.6: The Arduino Nano ESP.

Figure 17.7: The Arduino RP2040 Connect.

So the future looks very bright for the entire Arduino and microcontroller world in the processor area, at least. But it will most likely be a 3.3-volt world, as it seems everything new is focusing on 3.3 volts for power.

I can't wait to start incorporating E-Ink displays (Figure 17.8) into my projects. E-Ink displays are great from a low-power usage standpoint and they retain their display even when the power is off. All they need to do is get beyond the current 3 minutes between display update limitation and they'd be awesome for use in projects. I think it's just a matter of time before the E-Ink displays are ready to take over as the display of choice for Arduino projects.

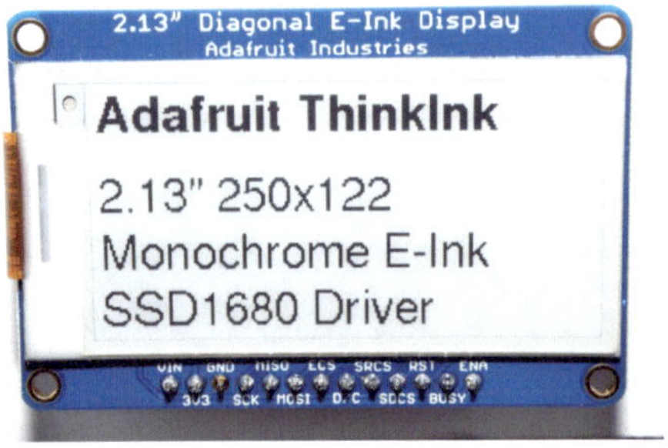

Figure 17.8: Adafruit 2.13" E-Ink display.

Voice recognition keeps improving over time. The new DFRobot Gravity: Offline Language Learning Voice Recognition Sensor (Figure 17.9) has 121 built-in fixed commands and 17 additional custom command words. The voice-controlled ham shack isn't far off at all now.

Figure 17.9:The DFRobot Voice Recognition Module.

Combined with Voice Recognition, the new DFRobot Gravity: GR10-30 Gesture Sensor (Figure 17.10) will allow you to create some pretty amazing projects that can be controlled using voice or hand gestures. Capable of recognizing 12 hand gestures, the possibilities for creating hand-gesture and voice-controlled projects are endless.

Figure 17.10: The DFRobot Gesture Sensor module.

Finally, what about radio-specific devices? Skyworks Solutions has the Si473x series of chips that are HF SSB and CW receivers on a chip. They've also brought back the Si4707 Weather Band FM receiver chip that includes the Specific Area Message Encoding (SAME) message encoding feature. I'd love to do an Arduino Weather Radio or HF Receiver project.

As for programming languages, that world is also starting to expand almost exponentially. No longer is it just the Arduino IDE's version of C and C++ to program with, now you have a host of other languages joining in on the fun.

Python, in the forms of MicroPython and CircuitPython is another very easy to learn and use programming language. Many of the recent microcontrollers, including the Raspberry Pi Pico, are designed for use with Python. Python on microcontrollers natively supports multi-core processors and multithreaded applications, so you can create some pretty cool multitasking programs with Python. Python can also be used interactively like BASIC, allowing for quick testing and debugging of your programs.

But there's other languages too. Standard versions of C and C++, MMBASIC and Piccolo Basic have been introduced into the Raspberry Pi's RP2040 processor world. Other languages are sure to follow. Several more recent Arduino boards, such as the NanoRP2040, Nano ESP32, Nano BLE, and the GIGA R1 now provide support for MicroPython.

Then there's the new Arduino IDEs. Arduino 2.0 is an update to the workstation-based Arduino IDE. IDE 2.0 has improved performance, an improved user interface, and many new features, including auto completion and a built-in debugger. IDE 2.0 also allows you to sync your sketches with the Arduino Cloud.

Arduino Cloud is Arduino's cloud-based platform for creating Internet of Things (IoT) projects. With Arduino Cloud, you can use your web browser as an Arduino IDE. This web-based version of the IDE allows you to work on your Arduino sketches from a variety of computers. Arduino Cloud also provides a "back-end" to transfer data between your devices and is essentially an IoT MQTT broker located on the Arduino servers.

And finally, the Arduino is at the heart of one of the newest ham radio-based hobbies, Picoballooning. Using something as simple as a Mylar party balloon or a larger ultra-lightweight high-altitude plastic balloon, hams are sending Arduino-powered experimental payloads (Figure 17.11) high into the sky, sometimes as high as 45,000 feet (15 kms) and tracking them using WSPR and APRS as they circumnavigate the globe. One recent Picoballoon designed to go straight up and back down was launched by Tom Medlin, W5KUB, and reached an altitude of over 109,000 feet (36.3 kms) before being safely recovered. Because of the highly accurate Arduino-powered telemetry, the recovery team was within 100 yards of the balloon when it touched down.

Figure 17-11: The solar-powered Skytracker Arduino-powered balloon payload by Bill Brown, WB8ELK (photo courtesy Bill Brown, WB8ELK).

So I'll say it yet again, with the Arduino, you're only limited by your imagination. For the amateur radio Arduino enthusiast, these are fun times and there are still so many projects left to build. For me, I think my biggest problem is deciding what to build next.

Index